Occupation: Angel

Books by Marguerite Cullman

NINETY DOZEN GLASSES
OCCUPATION: ANGEL

MARGUERITE CULLMAN

Occupation: Angel

W · W · NORTON & COMPANY · INC ·
NEW YORK

COPYRIGHT © 1963 BY MARGUERITE CULLMAN
FIRST EDITION

Library of Congress Catalog Card No. 63-10096

ALL RIGHTS RESERVED
Published simultaneously in the Dominion of
Canada by George J. McLeod Limited, Toronto

PRINTED IN THE UNITED STATES OF AMERICA
FOR THE PUBLISHERS BY THE VAIL-BALLOU PRESS
1 2 3 4 5 6 7 8 9

FOR
Howard

Contents

1 ·	The Beginning	11
2 ·	The First Play	26
3 ·	Something Goes Whack	38
4 ·	Flashback. . . .	58
5 ·	Push Aside Sorrow	74
6 ·	My Education	88
7 ·	Search for Identity	103
8 ·	Still on the Train	112
9 ·	Angels with Clipped Wings	124
10 ·	A Star in the Kitchen	141
11 ·	Readings . . . Auditions	159
12 ·	The F.B.I. and I	178
13 ·	Larceny and Hot Ice	192
14 ·	Pound Wise	206
15 ·	Take the Case of Mister Roberts	215
16 ·	Then and Now	232
	Appendix	245

Occupation: Angel

1

The Beginning

It occurred to me that the room would have made an excellent stage set: it had no architectural balance and an unexpected, supporting pillar stood in the center, but it did have character. Curtains were drawn across long casement windows; a French door led out to a garden-like terrace and far beyond were unseen trees bordering a quiet lake; low copper bowls were filled with fall flowers; the old fruitwood furniture was well polished, and bright tongues of orange flame lit a corner fireplace.

If the setting had been identified as a cottage in Wales it might have housed a family in very modest circumstances. Actually, it was a penthouse apartment just off Fifth Avenue in the Fifties, the lake was in Central Park and, as a way of living, it was an expensive whim.

Occupation: Angel

Across the room, I caught my husband's eyes. He had been following my obviously contented survey; he stretched and smiled.

As I look back upon the evening, it seems an unlikely moment for us to have embarked upon a strange career that would bring a return of a couple of million dollars.

Howard thought he heard the baby. I listened but there was a soft silence; not even the hum of traffic fifteen floors below intruded.

"I left her door ajar," I reassured him, "and we'd hear her . . ."

But Howard was off to see for himself. It must have been the same instinct that led him to telephone me twice a day, as if in leaving for the office he had abandoned me to frightful, unimaginable danger.

"She's *asleep!*" he announced when he came back, as if the baby had chosen the least likely and most brilliant way of using up the interval of time.

We settled back again. I looked critically at the square of needlepoint I was working. It was optimistically called *Fruit: Still Life*, but the plums resembled purple stones and a neatly dissected pear looked not only anatomical but nude and embarrassed at having its inner core perpetually exposed. Obviously, I had no talent for needlework. I would have preferred to read but I considered Howard temperamentally unsuited to such an arrangement. He is a glancer rather than a reader and on such rare, quiet evenings at home he tends to scan a newspaper and, halfway through any interesting item of news, stop and give me the benefit of his opinion on the subject in order to show me that, although otherwise occupied, he is happily aware of my presence.

I, on the other hand, read compulsively, and after five minutes I am lost to everyone and everything but my own private, immediate world. If interrupted, my responses either are

The Beginning

noticeable by their absence or are confined to a vague, resonant "um-hmmmmmmmm." It is a characteristic ill-suited to keeping warm an intimate family relationship.

I tried turning the needlepoint upside down to see if the womb-like pear improved with inversion. It didn't.

"This is a real dose of family life. Sure you're not bored?" Howard asked anxiously.

I smiled and shook my head.

"You don't miss having a job?"

Again I shook my head . . . this time with more vigor than honesty. The right to keep my job had been my only premarital request.

If anyone had asked me, before the wedding, if marriage and a career could be combined easily, I would have come out unequivocally in the affirmative. I even might have maintained that in my job as one of the editors of a national magazine, my time was more or less my own and I never worked after six o'clock. Fortunately, no one ever asked me, and my husband-to-be took for granted that if I wanted to keep on working, I was clever enough to manage.

He was wrong; and during the six months following the honeymoon I discovered that my all-through-at-six story was a myth. It was one thing to break a date with a beau on the reasonable grounds that something had gone wrong at the magazine and I'd have to stick with it, but quite another thing to ask your husband to explain the same problem to ten dinner guests.

The most surprising revelation was that houseworkers were not natural-born executives who could handle all emergencies but constantly looked for help and leadership from, of all people, me!

At the same time I found that I was accusing the laundry of shrinking my best linen tennis shorts quite unjustly and that what I needed to find was not a new laundry but a capable

Occupation: Angel

baby nurse who would be available in the early winter.

I resigned from the magazine and next to "Occupation" I mentally crossed off "Editor and Journalist" and neatly, if somewhat despairingly, lettered in "Housewife."

> Now mark me, how I will undo myself;
> I give this heavy weight from off my head
> And this unwieldy scepter from my hand,
> The pride of kingly sway from out my heart;
> With mine own tears I wash away my balm
> With mine own hands I give away my crown. . . .

Richard II and I were enjoying a new kinship. I had given up a job and not a crown but the essence seemed the same. I did not actually fool myself; I was fully aware of the fact that I had a tendency to dramatize. I did it as instinctively as water follows a river bed. I enjoyed my private exaggerations; they made my days and my nights seem slightly larger than life size.

Sometimes that summer I would lie completely immobile in the warm, summer sun and contemplate the thought that perhaps at that exact, seemingly slothful moment I was making an exquisite pair of hands. Michaelangelo's achievements paled in comparison. My approach to the whole subject of gestation was vague but romantic.

And after the baby was born, I looked upon her with wonder. It was as if I had written a masterpiece in my sleep, whole, complete and without need of rewriting or editing.

My life was settling into easy lines. I was happy; and yet, I suppose, years of work had formed too strong a pattern for me to be able to give it up without it leaving an empty space against my sky.

Howard looked up at me again across his sheaf of papers. The top page was filled with rows of figures. I had long since discovered that he read balance sheets with the avidity most

The Beginning

men reserve for murder mysteries.

"Look here," he said, waving the top sheet at me, "this is mighty interesting. A fellow was telling me that if anyone backed every show produced on Broadway in a year, even including all the failures, he'd wind up making money. And I've got a breakdown here . . ."

"How much money would that take?" I asked, making my periodic effort to sound companionable and intelligent about finances.

"Roughly speaking, three to four million dollars. But suppose someone backed *half* the shows . . . or put up half the money for half the shows . . . or 10 per cent. . . ." He was off in cheerful contemplation of black versus red figures and involved percentages.

"When you were on *Stage* Magazine and used to pick covers . . ."

"*Help* to decide on covers," I corrected in the interests of modesty and accuracy. "Sometimes the cover was based on the theme of a new play; sometimes it was a portrait of the star. We had to decide in advance which plays were going to be hits. So each member of the editorial staff used to read the plays and then individually vote on whether or not he thought it would be a success."

"How often were *you* right?" he asked.

This was a competitive aspect that I never had stopped to consider. I had no answer but I had discovered that my husband liked a definite answer to a question. To say I didn't know would only irritate him; besides, the urge to recall how good I was in my job was strong upon me.

"Maybe nine times out of ten?" I suggested tentatively. If he looked incredulous, I was prepared to substitute a more conservative estimate.

"Wonderful!" He accepted with a simple faith. "You've got a job again, on your own time and at home. We'll back plays!

Occupation: Angel

You read them, pick out the ones you like, and all you have to do is keep on being right nine times out of ten. It's terrific! In fact, even if you're right only eight times . . ."

"Dear God," I was praying quietly, "just get me out of this one and I promise not to embroider my stories any more. . . ."

Even if my boast had somehow clicked with the truth, I realized that the plays I passed on had been pre-sifted. They represented the probable giants of the season with star performers. Then too, my voice had been only one in four, a circumstance which made us all feel united and somehow exonerated in any misjudgment, but completely individual and prideful when we were right. At the moment I could see no easy avenue of escape from Howard's new idea but then, at the moment, there was no available play to consider. Time stood between me and a decision and diluted the threat.

I was undeniably attracted to the thought of backing; at the same time I mentally withdrew. It was not just nervousness at having the full responsibility of making the decisions, it was something deeper. Perhaps I was a little bit afraid of the theater.

For in my early childhood, the words "theater" and "sin" were, if not synonymous, at least closely related in my mind. And as it was an era of frenzied propriety, these words acted as sharp catalytic agents one upon the other, making sin more attractive and the theater more exciting.

My grandmother had defined what she felt was a reasonable code of behavior: the theater, like alcohol, could be taken with impunity in small doses, like a glass of sherry in mid-afternoon.

Hard liquor which fit into the "sin" department was almost never served in our house, but I knew from observing the highly moral cartoons in the *Evening Mail* what happened. In the newspaper drawings, children were obliged to don shawls or clutch their coat collars together, according to sex,

The Beginning

while they stood barefooted and shivering outside a saloon waiting for their father to come out. I considered this remarkably talented and shrewd of them, as once my father left the house in the morning I never had the slightest clue as to his whereabouts. He was "downtown" or, in the summer when we all went to the seashore, he spent five days a week "in the city." Fortunately, he showed no signs of taking to drink and I was, therefore, saved the probable humiliation of standing in front of Delmonico's when everyone else knew he was at Rector's.

Young as I was, I had the consequences of drink pretty well worked out in my mind; it was the evil of the theater that puzzled and fascinated me. And it came sharply into focus the summer I was six because it precipitated a scene between Mother and my grandmother, whose visit to us was drawing to a close.

I was in the corner of the room making a halfhearted effort to teach my puppy to sit down and give the paw and I was only vaguely aware of the fact that my mother had asked my grandmother what train she intended to take back to the city on Sunday. There wasn't much choice; there was one in the morning, one in the afternoon. Grandmother seemed uncertain as to her exact plans and wanted to know if it made any difference.

Mother cleared her throat and nervously dropped her voice to the pitch that alerts all inquisitive (snoopy) children to the fact that they are about to hear something they shouldn't. I immediately stopped talking to my puppy, the better to hear, but continued to make silent gestures of admonition to show how completely absorbed I was in my own world, at the same time catching a firm hold on my dog's collar so that he would not wander off and give me away.

"Mamma," Mother began in her giveaway tone, "you remember the party I told you about where we met the *actress*, Blanche Ring? Well, she's opening in a show that will try out

Occupation: Angel

in Long Branch, just a few miles away and . . . you know how fascinated Albert is with the theater . . . well, he, I mean, I . . ." Then it came out in a rush: "We invited her here for Sunday dinner!"

I cannot in honesty say that my grandmother stiffened, but there was an impression of reinforcement of the backbone and her lower lip achieved the almost impossible: it narrowed still further. Disapproval was implicit in the gesture but she refrained from making any comment. She was truly fond of my father and, if Albert were the instigator of the plan, she was not going to be trapped into criticizing him.

"But Mamma," my mother was answering the wordless look, "when *we* were children and the Barrymores took the house next door, you let me play on the beach with Ethel and Lionel all the time!"

To me this was an entirely new and fascinating aspect of my mother's past.

My grandmother had a fund of maxims against which it was useless to deploy verbal battalions. And when she brandished: "The *Barrymores* are the exception which proves the rule!" my mother retreated into silence. Then, apparently as a safeguard against contamination, Grandmother added: "If I take the afternoon train, I'll eat with the children."

Mother looked dreadfully upset and I suspected that she was sorry about the whole thing. Nevertheless, she plunged ahead:

"But Mamma, we were planning on having the children at the table too!"

Grandmother may not have approved of actresses but she certainly made a dramatic exit, tossing the final line over her shoulder:

"I have decided to take the morning train so there is nothing further to discuss . . . especially in front of a *child!*"

My mother cast a stricken look at me. My first impulse was

The Beginning

to comfort her but I followed a second and shrewder one. I made her speak to me twice before I could be roused from my all-absorbing world of dog training and be told to go outside and play.

My standing with my two older brothers rose immeasurably when I broke the news. Speculation as to how the actress would behave occupied us for the rest of the morning.

"Probably, she'll dance on the table," I suggested hopefully.

"And I get what-for if I even put an elbow on it!" The fat brother was rankling with the injustice of the world.

"She might," the thin brother said dreamily, "drink champagne out of her slipper."

"Father wouldn't *stand* for that," I said firmly. "You know how he is about germs!"

Just the same, I saw no harm in adding the *possibilities* of such lurid behavior as I called on my friends one by one in the little seaside resort and told them about the impending visit. I was in a mighty enviable position; I knew it and I savored every detail. Actually, I didn't *have* any details and so I was obliged to supply those out of inspiration.

One could not be sure of the exact moment of our guest's arrival, but nothing ever changed the one o'clock timing of Sunday dinner. This meant that we would be out on the porch precisely at two. My invitation to my friends was unspoken; so were the acceptances, but everyone planned to line up on our front lawn below the porch to observe the actress.

That Sunday morning I remembered to smooth down my skirt every time I sat back on the pew and to refrain from scuffing my white buckskin shoes. Just before the appointed hour, I was given a last-minute inspection and approval. My starched white dress was unwrinkled; a wide, black patent leather belt encircled the lower part of my behind and dipped gracefully in the front, giving the effect of a large, low stomach superimposed upon my slight frame. My short socks were

Occupation: Angel

pulled up and a chunky bow of bright, Roman-striped, silk hair ribbon had been tied at an unlikely angle on my soft, uncooperative hair. Sartorially, if not emotionally, I was set.

So far as we children were concerned, the meeting and the dinner that followed were a shocking failure! The guest of honor was quiet, soft-spoken, not as pretty as our own mother and was older; she was indistinguishable from the other ladies. For a while, I entertained the hope that this was not an actress at all but an imposter. I thought it suspicious that she used two names, Blanche Ring and Mrs. Charles Winninger. Then, with a reluctant concession to the workings of Life, I gave up. By the time dinner was over I was not even surprised that she had not pranced on the table or drunk from her slipper. Her conversation appeared to interest the grownups but left us cold, and there had not been so much as a single admonitory *"Schweigen, die kinder!"* from Mother. In our home, German was not so much an accomplishment or a link with another people as a tool for bypassing the children. And when Father, carried away by the charm of the story he was telling or by the violence of his feelings about politics, stepped beyond the lines of protocol, Mother would curb him (and thus alert us that something forbidden had just been said) with *"Schweigen."* The fact that the admonition had not once been used was final proof of the dullness of the dinner.

Not only were my own dreams shattered, but in a few minutes I would have to meet the disappointment and then the derision of my friends!

I left the dining room reluctantly. Mother was ahead and was the first to see the five children lined up on the the lawn, solemnly staring at us. She caught on, intuitively. Swiftly she bent down and whispered that I was to take them around to the back to play.

"We do *not* show off our guests," she said. "We don't even *discuss* them!"

The Beginning

I saw a ray of hope. "We don't?"

"Definitely not!"

The children gathered around me in a circle of scrubby pine which was considered a sheltering thicket.

"Which one was the actress?" the boy next door wanted to know. He was a wretchedly practical child.

"Why the one in lilac, of course! . . . but I shouldn't tell."

"Why not?"

"Because it isn't polite," I said, smugly. I could see that etiquette had its own possibilities.

"Well, what did she *do*?" he persisted.

It must have been then that I discovered that nothing so thoroughly perverts the truth as telling only part of it.

"I've just been forbidden to discuss it!"

"Jiminy cricket!" he breathed in awesome respect. "It must'a been something! Wish I could'a seen it."

Time telescopes our memory and so I do not know if it was just after this event or later the same summer that Father confirmed the sense of evil connected with the theater by entering into a bargain to take me to a show and then dragging me out after the second act.

For some reason which I cannot recall, Father was spending a few days during the week at the seashore and, as usual under such circumstances, he was restless.

"Why don't you take Marguerite out for the afternoon?" my mother suggested brightly. "Perhaps just the two of you could go for a drive."

Father and I looked at each other. Togetherness had not hit the country and the prospect of a long afternoon together filled us with mutual doubt. Still, a ride was a ride. . . .

I took along a navy blue reefer coat to guard against ocean breezes and Father and I stepped jauntily into the pearl gray Roamer which one entered from the *side* as opposed to the old car which one entered by climbing up two steps in the

Occupation: Angel

center *back*. We headed along the ocean drive in the direction of Long Branch.

"What do you say to our seeing a matinee? We'll keep it as a kind of secret between the two of us."

His tone was lighthearted but it didn't fool me. The pact of silence was necessary to keep the rest of the family unaware of our dallyings with evil. I had heard it said that my father was an interesting man and now, staring at him with new-found admiration, I concurred. As we walked into the theater we held each other's hand. . . .

The first act passed in a comfortable haze. I couldn't follow the plot very well, but I was content to watch the actors come and go and listen to the cadence of their voices. There was a big house party and somebody's pearls were stolen; everyone was upset. I felt cheerful.

At intermission we went into the lobby and a gentleman greeted my father: "Picking them a little young, aren't you?"

Father smiled his agreement and told me to tell the gentleman my name. I threw in my age for good measure. They continued to chat for a few moments; then the gentleman gave me a thoughtful look and said: "This show isn't exactly *The Bluebird*, y'know, isn't six a trifle early for this sort of thing?"

"It's a perfect age," Father pointed out. I was warmed by his quick defense of me until he added: "She hasn't the slightest notion of what it's all about."

I was wounded. I wanted to protest that I did so know what it was all about. But I had learned that when grownups talked about you as if you weren't there, it was better if you too pretended. Besides, I knew I had *not* understood very well.

Father, I determined, would be proud of me from there on. And I followed every word of the second act with careful absorption.

The second act went right on with the story just as if every-

The Beginning

one had stood frozen while the curtain was down. Almost everyone on stage seemed to think that the pearls had been taken by the nice young man. I didn't agree with them. And the leading lady liked him and she didn't think that he had stolen them either so that made two of us. She thought the fellow with the mustache had 'em. And that night after dinner while everyone else was downstairs, she went up to search his room. That was very exciting. She kept looking over her shoulder and listening but finally when the door *did* open, she never heard it at all! The mustached fellow closed the door behind him and locked it. Well, instead of being mad at her for suspecting him, he went right over, said "a-ha!" and tried to kiss her. But she didn't like him and she ran around a table and he ran after her. Neither of them ran very fast but he was panting something awful, so they gave up and just grabbed hold of the table from opposite sides and stared at each other. He said "ha" a couple of more times and then started over very sneaky-like to grab her. I decided he was really mad, after all, and maybe he was going to choke her. She told him to stay right where he was or she'd scream and she took a deep breath to show she meant it. Then he just smiled a sneery smile and said to go ahead, but that when everyone rushed up they would find her in his bedroom at midnight with the door locked. Well, this was the silliest argument I *ever* had heard because naturally everybody must know what time it was and whose room it was. But it stopped her dead in her tracks. She stared at him very hard and then *she* began to pant and clench her fists and then she whispered in a very loud whisper: "Trapped!" and the curtain came down on the second act.

I couldn't wait to show how much attention I'd been paying.

"Boy, is she in a spot!" I observed. I knew it was all right to talk because there was no one on the stage. Then I went on to say that I thought she should scream anyhow as I'd rather have all the other people mad at me than stay locked up with

Occupation: Angel

a sneaky fella like that! Well, a couple of people on each side of us looked at me very curiously. I figured that *they* hadn't been paying too much attention that act and they were interested to hear about it from me. In the middle of my explanation, Father grabbed me firmly by the hand and pulled me up the aisle and right out of the theater. And he said: "Good God!" although Mother didn't like him to say that.

"*Schweigen!*" I said, almost automatically in her absence.

Well, naturally, I cried at missing the last act and never knowing for sure who did have the pearls.

After we both calmed down, Father was nice about it all. He said that he had it on very good authority that the nice young man came in and fixed up everything: found the pearls, sent the bad one to prison and married the girl. And the reason he had left so abruptly was because he was afraid that if we waited until the end it would be too late to go to Huyler's for an ice cream and he was just dying for some. Oddly enough, he just drank coffee but I had a scalloped, thick glass dish with two scoops of chocolate.

* * *

My time sequences were confused. Going back twenty-odd years had been as quick and simple as taking two steps down. I had felt myself a child again, the child that was mother to my womanhood. And my blonde, warm, beautiful Mother was alive again, waging her endless rebellion against the righteousness of Grandma.

It took more than two steps to return to the present. And they were giant steps leading me to the Now; the Now when Grandma was a soft memory and Mother a recent wound; the Now when *I* was Mother and there was no one behind me on whom to lean.

The baby sucked at the bottle eagerly, her face flushed and moist from the effort, eyes half closed in contentment; long

The Beginning

black eyelashes curving down to her cheek. I tightened my arm around her and the fan of lashes flew upward. We eyed one another as if staring into twin mirrors of Today and Tomorrow. By her birth she had crowned me Mother, and almost before I learned my role she would reach out to retrieve the crown for herself.

"What are you two staring at?" Howard wanted to know.

"Ourselves," I whispered. "Isn't it weird; all of a sudden she'll be me and I'll be my grandmother—disapproving of something?"

"As you're in your twenties and she's still an infant," Howard said, "I don't think the problem is exactly imminent."

The First Play

The next morning Howard telephoned me from the office; obviously he had plunged promptly into our new enterprise.

"We have a play!" he announced and then proceeded to fill in some of the details. Our mutual friend, John Byram, who was head of the play department of Paramount Pictures, was an intimate friend of a successful playwright who had just finished a new play. We were to be given the chance to invest in it. Johnny offered to mail the play to us but Howard couldn't wait; he'd have the play picked up and rushed right up to me.

I was in no hurry to meet the test of my talent. I read the play carefully and concluded that God had decided to cooperate with me.

"I don't think it's good enough," I said happily.

The First Play

When pinned down to specific reasons, I suggested that it was too wordy and not very tightly knit.

Howard was shocked. "You can't say that to a successful playwright," he protested. "This is the man who wrote *The Barker* and *Sailor Beware*, a couple of the biggest hits on Broadway. Why, I'll bet he'd be so indignant he wouldn't let us put up the money."

"Well, don't tell him the reason," I countered. "Just say 'we decided not to make our debuts!' "

This had become a family gag and was based on an incident that had occurred at Bonwit Teller and was always good for restoring good humor:

The buyer of the Misses Dress Department had received an unexpected visit one Saturday from an elderly executive whose interest in young ladies was undimmed by his years.

"Miss D," he said, "I'd like to see a very smart evening dress . . . size twelve." And then, as if the request required some explanation, he added, "It's for the daughter of a friend —she's going to make her debut."

Miss D's suspicions were allayed and she trotted out white satins with shirred skirts, frocks caught up with girlish gardenias, and even a hoop skirted tulle, but the executive shook his head impatiently. It soon became apparent that what he had in mind was a tight sheath, preferably in black satin. "She's a sort of sophisticated debutante," he explained lamely.

Miss D produced just the number he had in mind. He threw it over his arm, jauntily waving aside her offer to have it delivered or even wrapped. "You can charge it back to my account . . . at the wholesale price, of course."

Miss D had a private laugh at herself. "The old goat! And for a few minutes he had me fooled. Daughter of a friend, indeed!"

But he returned the first thing Monday morning with the dress over his arm and told her to return it to stock, it wouldn't do. Confusion set in again. What, Miss D said to herself, if he had been telling the truth?

"Wouldn't you like to reconsider that white one with the little cap sleeves . . . ?" She pleaded. "It's so young, so fresh."

He shook his head moodily.

"Or I could shop the market this morning for something else . . . ?"

Abruptly he closed the subject: "I don't want a dress, *any* dress. The young lady—has decided not to make her debut!"

The talisman had its usual effect and Howard laughed.

The next evening he came home with a glint in his eye and a faint fragrance of dry martinis on his breath.

"I met Ken Nicholson this afternoon."

"Who?"

"Ken-yon Nich-ol-son," he repeated slowly and carefully. "The author of *June Night*."

"Oh, yes! Of course," I apologized and slipped my mind into its right slot.

"Well, as I was saying, we had drinks together and he's absolutely charming, and very intelligent; he teaches at Columbia University. You two are going to like each other," he added contentedly.

"Even after what I said about his play? Or did you tell him that I had decided not to make my debut?"

Howard smiled briefly, absently at my small quip.

"Did you know," he said ignoring my question, "that in financing a play, when you put up half the money you don't get half, but only *one quarter* of the profits? Interesting, isn't it?"

"Very," I agreed and knew that we had made our first investment.

But the percentage that we were to receive of the profits was the last thing in the world that we had to worry about just then.

* * *

The First Play

I was curious about the financial mechanics of producing. I was told that we were lucky; this would be an inexpensive show to produce and probably would be financed for about $16,000. This would be supplied by the backers. When I asked how much the producer put up, my theatrical mentor frowned. The producer, he explained, puts up his experience and his time against our money. And when the profits came rolling in, the first money would go to pay off the backers' entire investment and from there on we would share fifty-fifty with the producer. This sounded to me reasonable, even generous. I also learned that the money taken in at the box office for the sale of tickets was known as the gross receipts. The weekly expenses for the running of the show (theater rental, actors' salaries, etc.) was called, in theatrical parlance, "the nut," and when you deducted this from the gross the balance was the profit.

I realized that this was the roughest sort of sketch, but it seemed rather tidy and sensible and actually it was about as far as I cared to delve at the time. Then I thought of the sizeable pile that 16,000 crisp, new one-dollar bills would make and I wanted to know where it would go.

"That depends," I was told.

Finally he conceded that it *might* go something like this:

Scenery, Props and Costumes (which were almost nonexistent in this play)	$3,500
Designer's fee	750
Electrical equipment (rental)	500
Director's fee	1,000
Rehearsal expenses	2,000
Advertising advance	1,000
Miscellaneous	1,000
Legal fees	250
	$10,000

An additional $6,000 would go for what my instructor blithely called "recoverables." What he failed to tell me, and may not himself have realized, was that they were recoverable only in the sense that if the play was successful they could be charged off to weekly running expenses but if the play failed, they too were gone.

The Producer's Bond would assure Equity that the actors would have two weeks' salary guaranteed and that the stagehands would get similar protection. Another two thousand would be posted as a guarantee or deposit on the theater. (I thought that seemed shockingly high. Later, when *we* owned a theater, I realized that twice the sum barely covered expenses. Strange, how one's perspective changes.) Five hundred would go to the author as an advance; the balance, an indefinite sum, would come under the heading of Reserve Fund.

On the whole, it all sounded well planned and efficient to me. I just wished that I liked the play a little better.

* * *

In January the play had its out-of-town opening in Philadelphia and we went there. I was far from sanguine.

"*June Night:* A January Frost!" one headline ran. For once, there was a complete solidarity of opinion. As Howard read another of the criticisms aloud, ". . . over-wordy and loosely knit construction," a beam of husbandly delight spread over his face. "Practically your very words!" he said with renewed faith.

"There's no use trying to bring the show into New York in its present shape. Ken says he'd like to take his time, rewrite, maybe recast and open in the fall. Of course, it will mean refinancing. . . ." Howard was sounding brisk and professional.

In the fall, the New York critics saw eye to eye with their Philadelphia confreres. The play lasted for two days and there seemed to be a general feeling that it had had an overextended run.

To my horror, I found that we had subscribed to over 90

The First Play

per cent of the financing and we had dropped $23,000 out of the $25,000 which the two productions cost. But in the meantime we had become involved with two of the great hits of the season: *Abe Lincoln in Illinois* and *Knickerbocker Holiday*. We barely had time to recognize failure before we knew success.

John Wharton, the theatrical expert in our family's law firm, had come to us with an interesting proposition: Five famous playwrights, who were described as fugitives from the Theater Guild, had decided to form their own producing firm. For them, there would be no more scrounging around for the right producers, no broken promises, no excessive commercialism. All five would read each other's plays and thus derive the benefit of expert criticism and counsel. They decided to capitalize for $100,000; $50,000 would come from a group of select backers, the other $50,000 they would put up personally and thus profit in the dual roles of producers and backers. It was a tempting proposition for everyone concerned.

As backers, we would have the privilege of being in on all the plays written by Maxwell Anderson, Sam Behrman, Sidney Howard, Elmer Rice and Robert Sherwood. It was a backer's dream of exclusive participation. We felt strangely honored, as if we just had been proposed for membership in the Athenaeum Club in London. We put up $5,000, which was all that was open of the outside $50,000.

Almost before the ink was dry on our check, the Playwrights' Company gave birth to its first offering: *Abe Lincoln in Illinois* by Robert Sherwood, staged by Elmer Rice and starring Raymond Massey. We were pleased and excited and, unlike our first venture, I was in love with the play as I read it.

A week before it was to open, we went out of town. Howard's secretary telephoned to relay a few messages. I took the call. She wound up with: "Oh, yes, and there's a telegram for you and Mr. Cullman saying that after the opening of *Abe*

Occupation: Angel

Lincoln in Illinois there will be a supper party at the Elbow Room in honor of Raymond Massey, given by the backers."

"Damned funny system," I reported to my husband. "We're *told* that we're giving a party! Not asked if we'd like to, mind you, just *told!*"

"I guess it's the custom." Howard took it philosophically.

"I'm sure you're right," I agreed. "But I wonder how they'll divide up the cost. You know we have ten guests for that night. And although it will be great fun for them—and for us—I don't want the other backers to think we're being piggish."

Howard brushed away my concern.

That opening night was one of the really glamorous ones. The mounted police were lined up, holding back the strange, heterogeneous crowd that pushes and lurches, straining for a glimpse of famous figures. It was a deep, October night but the violet of darkness was driven back by neon lights and dazzling flashbulbs. A long line of limousines moved along with the jerky undulations of a mechanical snake and, each time a Hollywood star would emerge, a roar of approval went up from the crowd. The regular firstnighters were still strangers to us but they, the elect, seemed to know *everyone;* they smiled, waved and occasionally gasped each other's first names with such enthusiastic astonishment that one could only infer that this was the least likely place in which they might expect to encounter each other.

Our guests also were strangers to this world, and they took it as big as we did. Howard didn't even bother to keep the pride out of his voice when he told them that we were giving a supper party after the show.

"To meet Mr. Lincoln . . . I mean, Mr. Massey," I whispered. But it was too late. I had confirmed my own suspicion! And to this day I keep the awesome feeling that Raymond Massey *is* President Lincoln and that, consequently, the whole story of the assassination must be highly apocryphal.

The First Play

Intermission was a facsimile of the arrival scene except that the big stars who felt harassed by crowds stayed in their seats while the ones who relished adulation gave their autographs in books, on programs, on scraps of paper. And they were not all teenagers who pleaded for these signatures; sometimes they were the raggle-taggle Broadway bums of both sexes; sometimes they looked like glamour-starved housewives. I wondered if it were the same crowd regrouped, or if the first had departed and sent in a scrub team.

Howard and I had come from entirely different backgrounds and we still did not know all of each other's friends. I was pleased to discover that I knew some of the audience; Howard seemed to know three to my one. And then a mutual acquaintance whom we both knew very slightly, greeted us warmly: "Well, it looks as if you have a big hit in this!"

I couldn't imagine how he knew that we had an investment in the play; no one had told me about the Broadway underground.

A few minutes later, someone else clapped my husband on the back and said: "I hope you have at least 50 per cent of this . . . it'll run forever."

I tried to explain that we had a bare one-twentieth, but I might just as well have been speaking Hindustani. "Well, lots of luck! I wish *I* had the other half!" he continued jovially.

It took me quite a while to realize that it's almost an unwritten rule: on opening night no one ever listens to anyone else!

A third chap spoke to us and we never did identify him; Howard thought he was a friend of mine and I thought he was a friend of his. In any case, he too seemed to know we had a vested interest in the play.

"You sure know how to pick 'em!" he said.

I didn't try to explain how the Playwrights' Company worked and how little we really had to do with it but, just

Occupation: Angel

vaguely, to keep the record straight, I said something about our score being tied by the previous night's failure. He looked surprised and embarrassed; I gathered that I had committed a solecism and that one does not mention a flop! Backers must take their cues from the sun dial.

When we arrived at the Elbow Room after the show, we checked our wraps and joined the long, curved reception line. We were early but the group was lining up behind us rapidly; we could see both ends.

Howard exchanged a flash of recognition with someone at the head of the line.

"That's George Backer, an old friend. I don't know whether you've ever met him . . . he once ran for Congress . . . and that's his wife, Dorothy, next to him . . . they own and run the *Evening Post*. . . ." Howard was rambling on informatively, giving me all the information I needed to know except the one great awful truth which was just dawning on me: Standing at the head of the reception line next to Raymond Massey were a Mr. and Mrs. Backer . . . "The Backers!"

A moment later we were there! We could neither apologize nor explain at that moment. I could only shake hands with my host and hostess and introduce them to our ten guests who seemed meanwhile to have split like amoebae into an infinite and agonizing number.

For a long while after that, if anyone asked us on an opening night: "Are you backers?" we'd have the instinctive impulse to protest: "Oh, no, we're the Cullmans. The Backers are over on the left, two aisles ahead!"

Two other plays in which we had no interest opened that week and quickly were branded failures. We were ungenerous enough—and naïve enough—to be pleased. But that was before we knew that plays are not wholly separate entities but rather part of an interlocking chain known as Broadway and, the more successful the season in general, the more interest

The First Play

is generated and the greater the chance for individual acceptance.

That same week, the third play in which we had an investment opened: *Knickerbocker Holiday*, a musical comedy by Maxwell Anderson with music by Kurt Weill. Once again, the Playwrights had come up with a hit; again they wisely had chosen Jo Mielziner for the settings; Josh Logan staged it; Ray Middleton sang one of the leading roles, Richard Kollmar another, but the real star of the show was Walter Huston, a man who could not sing . . . or, at least, who hadn't. Broadway was quite agog over the casting. But, fortunately, no one ever told Huston he could not sing and when he appealed to the young girl of his choice to share his "last few golden days" in the "September Song," the audience did everything but stand up and cheer. There were those who said later that he did sing it . . . sort of; there were those who said he just recited it . . . musically; and there were those who said that he was carried along by Weill's music which was warm, rich and appealing. In any case, there was magic in the air and that night two new images were born: the musical star without a voice; and an older man in love with a young girl became, instead of a buffoon, a romantic figure.

I was enjoying our closer contact with the theater but I wasn't, like Howard, taking any real personal satisfaction from it. The Producing Company graciously allowed me to read the plays prior to production but I had no cause to make any comment. We had bought a ticket on a merry-go-round of success where we could neither call the tunes nor the stops. It didn't matter; each time round everyone pulled a gold ring.

I read the next Playwrights offering, *American Landscape*, by Elmer Rice.

"I don't like it!" I said. "It's fantasy that never gets off the ground!"

"Whoa!" Howard cautioned. "Why, this is the man who

Occupation: Angel

wrote . . ." His voice trailed off as he began to recognize a pattern.

"Uh, huh," I confirmed, "he did. But maybe they all write turkeys once in a while."

"Well, supposing that you're right; there's nothing that *we* can do about it. We're committed to *everything* they do. Anyhow," he added, "maybe *you* don't like it but apparently, Sherwood, Anderson, Howard and Behrman *do!*"

"Maybe they do . . . but maybe," I said thoughtfully, "they don't like it any better than *I* do! As you say, we are committed; perhaps they, too, are committed. Two out of the five have just had their plays produced and now Elmer Rice says it's his turn. His play is ready—it isn't bad, it just isn't very good. Who's going to tell him? Who's going to bell the cat?"

No one—certainly not the Playwrights—ever did decide.

The play was not a success. It closed after five weeks.

Elmer Rice went right back to work on a new one.

The fourth offering of the Playwrights was S. N. Behrman's *No Time for Comedy* and everyone agreed that this was a play that had just about everything. It also had just about "everyone who was anyone" in the cast. Katharine Cornell played the lead—her first major comedy role—and the cast included Laurence Olivier, Margalo Gillmore, John Williams and Robert Flemyng. This was the kind of casting of which playwrights, producers and backers dream. And judging by the audience response, it was their dream, too. All of the young wives in town tried to dress like Miss Cornell, look like her and adopt her sophisticated charm, worldliness and understanding; all the young men tried to imitate Olivier's accent, his crisp dialogue and fatal appeal. Unfortunately most of them lacked Valentina's designs, certain endowments of Nature, the accident of English birth and having Behrman write their lines. But *No Time for Comedy* left its mark on the

The First Play

1938-39 season which was shortly coming to an end.

The theater, scorning both the Roman and the Gregorian calendars, arbitrarily counts its year as beginning in the early fall and closing in June. Let others see the old year out at the end of December; the theater season dies in June and is not reborn until September.

That season Broadway scored eighty shows—musical and dramatic—and the season adhered strictly to the traditional rating of 80 per cent of failures to 20 per cent of successes, or four to one on the debit side.

We had participated to the extent of five of those shows and somehow had managed almost to reverse the over-all percentages.

3

Something Goes Whack

As it proved impractical to follow my husband, shaking my head and making Indian signs of negation every time he said we were looking for plays in which to invest, we eventually found ourselves with another manuscript. And this, once again, would require a personal decision. Perhaps, I said to myself, it won't be any good. I think I almost hoped that in full conscience I could turn it down.

"I like it," I admitted to my husband in the embarrassed tone of one who is displaying to the horsiest set in Virginia her obviously spavined, wind-broken old mare. "Of course, you couldn't exactly say it had much of a plot, just vignettes of family life, and nothing really happens except that in the end, the husband gets baptized."

Something Goes Whack

Howard agreed that that didn't sound like too gripping a situation.

"Are you sure you want to go in?" he asked.

I was sure. In fact, I was insistent.

"It hits something inside of you; and it plays right . . . in my mind's eye," I explained. But this seemed to muddy rather than clarify, so I switched to a financial aspect and pointed out that it would be a very inexpensive show to produce since it had only one set. But I am afraid that I also suggested that regardless of how much money might be open, perhaps something like 20 rather than the 93 per cent we had put up before might be reasonable; after all this was just a little thing that I had taken a fancy to and there was no need to be reckless.

The two authors and the producer turned out to be equally cautious fellows who suggested trying out the play in Skowhegan, Maine, in the summer theater so that all of us could get a look at it. Then, if the play didn't look sufficiently promising, we'd skip a New York production and drop only half of the money.

And so we put up $5,000 out of the $25,000 for *Life With Father* which had a longer run on Broadway than any other play in the history of the American theater, was sold to the movies for a fabulous sum and yielded a gross return of well over $10,000,000!

Some plays, regardless of their basic merit, seem fated from the start for failure, some for success. This was one of the latter. And most of the things that happened to it were without plan or schedule.

Initially, Howard Lindsay had no intention of playing the part of Father but neither the producer, Oscar Serlin, nor the two authors, Howard Lindsay and Russel Crouse, could find anyone whom they considered ideal for the leads. So Howard Lindsay decided that he and his wife, Dorothy Stickney, would do it in Maine; plenty of time for the final casting if it came

Occupation: Angel

to Broadway.

They were a little worried about the first act but very confident of the second.

Opening night in Maine was a surprise; the first act proved a riot, but the second one slumped. They studied the script; they watched the audience reaction, Crouse from the back of the house, Lindsay from the stage as he portrayed Father. And they came to the conclusion that their timing was wrong. They had to move from scene to scene more rapidly. They cut six lines here, four there; when they were finished pruning, the action was faster, the big laughs were closer together, they had a solid second act and a great prospect for Broadway. Also, there was no longer a question about anyone else playing Father and Mother.

Word had gone around that a hit was coming in, but on opening night all of us were tense and excited.

In the first act, Father's gruffness and impatience with the timid maid got a big laugh.

How well, I said to myself, she gives the impression of nervousness even though she does practically nothing . . . and then, whoops! That was a new bit of business: the maid dropped the entire tray of breakfast dishes, threw up her hands and dashed off.

The audience howled with delight. Mother made a gesture of weary resignation and played the rest of the scene half up, half down as she picked up halves of grapefruit and cleared the debris. Not until three hours later did I find out that the whole thing was truly an unrehearsed accident. In the last scene one of the boys failed to show up on cue but the action went along smoothly without him. He was in his dressing room happily reading his telegrams of congratulations and oblivious of the fact that the play was still going on. Only the thunderous applause roused him in time for the curtain calls.

The success of *Life With Father* renewed my husband's

Something Goes Whack

faith in my ability to judge plays in manuscript form. I myself wasn't so sure. I knew that Howard possessed a kind of chauvinism that extends all the way down the line from our country and our government (whichever party happens to be in power) to our family and all its possessions. Other people's children left the house "a mess"; ours gave it "a lived-in look." Even the beautiful Gordon Setter we got from England that turned out to be that rarity among dogs, a half wit, was automatically credited with sagacity because he was ours, wasn't he? And so I knew that, right or wrong, I'd be protected in my decisions. But I began to feel a deep, secret sense of security and a confidence that I knew what I was doing. The decisions came spontaneously; there was even a faint echo of familiarity in my words, as if I had said the same things before long ago, in some other life.

I read half a dozen outside plays and turned thumbs down on all of them. The producers must have gotten similar, poor reactions because they dropped their options on three of them; three were brought to Broadway but closed almost immediately.

Hard on the heels of *Life With Father* came two more offerings from the Playwrights: *Key Largo* by Maxwell Anderson and *Two on an Island* by Elmer Rice. Both had respectable but unspectacular runs of three months.

Meanwhile, the Playwrights had been diminished by one. In the prime of his writing career, Sidney Howard had been killed in a sad and senseless accident.

The theater season of 1939–40 finished in a blaze of glory with another Playwrights' production: *There Shall Be No Night* by Robert Sherwood.

Mr. Sherwood had heard a broadcast from Helsinki which inspired him to try to dramatize the betrayal of Finland. Although he was in the middle of work on another play, he dropped everything and worked intensely, feverishly on his new

Occupation: Angel

idea. As he wrote, his characters seemed to him to be Alfred Lunt and Lynn Fontanne. They had just finished an exhausting, twenty-thousand-mile tour of the country and were leaving for a long-delayed vacation on their farm in Wisconsin when Mr. Sherwood gave them his completed play to read and to consider for the following season. Before their train reached Chicago, Bob Sherwood had a wire from them saying that they would be ready to go into rehearsal just as soon as they could turn around and get back.

Both Mr. Lunt and Miss Fontanne had been associated closely with the Theater Guild for many years and so the production came out under the joint auspices of the Playwrights and the Theater Guild. There was glory enough to share.

For us, the following year in the theater, the season of 1940-41, was an exciting one both from the standpoint of general dramatic enjoyment and because our own career as "angels" was soaring. We had made lucky or wise decisions and it had become apparent that we were not investing in the theater because Howard was intrigued with an actress or even for the more respectable and domestic urge to find a place for a stage struck son or daughter. We were interested. It was as simple as that!

Producers who never had taken outside backers, preferring to finance their plays personally, began to consider splitting the risk; after all the cost of production was going up and $75,000 to $100,000 was a lot of money for them to invest in one musical. Quite a few doors began to open to us, including Sam Harris'.

Moss Hart, as anyone who has read *Act One* knows, had achieved success in collaboration with George Kaufman. But into every collaboration, much as into every life, comes a period of uncertainty, of doubt. This is what I am in consanguinity, what am I standing utterly alone? And this Moss Hart had to know. The story of *Lady in the Dark* was the result of an experience Moss Hart himself had had in the office of a psychoanalyst; at first he thought of it as a straight play with possibly Katharine Cornell in the lead, but as it developed the dream sequences demanded music and a different treatment. Gradually the image of Miss Cornell softened, blurred and was transformed into Gertrude Lawrence. It took a bit of negotiating to persuade Miss Lawrence to abandon the straight drama which she loved and return to a medium she had put behind her. Kurt Weill wrote the music, Ira Gershwin the lyrics and Sam Harris was prepared to produce the show. But it was going to be expensive, at least by the standards of those days and this time Sam Harris was willing to take outside money.

Four or five of us gathered in Mr. Harris' office; Moss Hart rode down with us, carrying his play under his arm in a manila envelope; he was going to read it aloud to us and he said that he was nervous. He didn't appear to be concerned in the slightest; later when I knew him better I discovered that the more poised and at ease he appeared, the greater the chance that he was inwardly in agony.

He pulled up a straight chair and quietly, almost impersonally, began to read the directions for the sets and on into the story. His voice was quite lovely but it quickly faded from my consciousness. I was unaware of everything except the action which seemed to be unfolding there in the office before my eyes.

In one scene the lead, Liza, recalls an event in her childhood: There were a group of people in evening clothes; the woman of great beauty in the center is Liza's mother. Her father walks in with the four-year-old Liza in his arms. He has brought her to meet the guests and to say good night. The guests make the usual polite fuss over the child and her looks, but her father laughs them off and protests that he is reconciled to having a plain-looking child. "One beauty in the family is enough, I can tell you," he says, smiling at his wife. "I couldn't

Occupation: Angel

stand two. Daddy's little ugly duckling, isn't she? Come, Liza! Sing us your song, and then a good-night kiss." But Liza buries her face in his shoulder, then bursts into tears and runs from the room.

"I ran to the nursery and looked in the mirror," the grown-up Liza recalls. "I felt ugly and ashamed. When my mother came in, I hated her because she was so beautiful!"

At this point I felt that I was hearing two stories: mine and Liza's.

I was so used to making the decisions about plays that I had rather forgotten Howard's presence, otherwise I might have worried over his reaction to the play.

"All—this—throws me off," Liza says as she walks into the office. "I know very little about psychoanalysis, Dr. Brooks, but I do feel there should be a beard and a Viennese accent around someplace."

The slightly mocking rebellious quality, I felt would put most of the audience on the heroine's side. Especially when she says: "I have nothing but contempt for women who spend their days pouring out their frustrations at so much per hour." And then adds: "I don't particularly believe in psychoanalysis."

By the time the play was over and the neurotic heroine had begun to understand herself and had found the right man, I felt as if I just had undergone a successful analysis myself. (Even before I had heard the haunting and lovely music, I was completely sold.)

"Do you *really* like it?" Moss asked.

My first reaction was almost one of indignation. Of course I "really" liked it. Who could fail to feel that way? The question itself seemed a pose. Then I took a closer look at Moss. At the moment, he was not a theater-wise man judging a vehicle; he was half of a successful team suddenly on his own; he was an author getting an audience reaction. I think I never

Something Goes Whack

saw such vulnerability in a man's eyes. I don't recall the words in which I told him how much I loved the play, all I remember was that there was warmth and understanding between us.

Then we indicated to Sam Harris our interest in participating, subject to Howard checking the contract and the business aspects.

"To what extent do you want to go in?" Howard asked me.

"All the way! Whatever is open," I said and then wondered at my own boldness.

"What's happened to Miss Pennypincher?" Howard teased.

All semblance of reluctance had gone. I was committed to the theater.

On the way home, Howard got down to business. "It looks like a $100,000 production; Mr. Harris would like to accept half of that as outside financing, preferably from one or two people without it being shopped around. Fifty thousand is a little steep—even if you do seem crazy about it. Marshall Field has been asking me to cut him in if I find anything we really like—maybe he'd like to split this with us."

I nodded my agreement. Yes, I really did like it that much and yes I thought it would be fun for Marshall.

"You really do think the audience will go for all that analysis stuff? I feel like one of the characters in the play who said: 'God knows I don't know anything about it but you do hear the damnedest things! One person discovers he was frightened by cornflakes as a child. . . .'" Howard wanted reassurance.

"Did you think that childhood business rang true?" he pursued.

"The story of my life," I answered flippantly, because it really had touched me.

"Oh, come now!" Howard protested. "To begin with, *you're* pretty!"

That is a statement to which no woman, unless she is an

Occupation: Angel

utter fool, would take exception.

"And," he continued, "you and your mother adored each other."

"Nevertheless," I said, "because of just such an incident, I never really understood her . . . or her feelings about me until after she died; and then it was just by accident. . . ."

The incident happened twenty-five years earlier, when I was a year or two older than the child in this play. . . . It wasn't shocking at the time; no scenes, no tears, and yet I can remember it almost word for word. I can even remember the position in which the two lady callers were sitting in our parlor. They had come calling a little earlier than they were expected and my mother was not yet home.

My nurse washed my hands and sent me in to bridge the gap. The ladies greeted me effusively and each gave me a parched kiss that smelled of stale powder. They were sisters, older than my mother, elaborately dressed in the fashion of the times and their faces appeared to have been much harried in an effort to achieve a beauty that they did not possess. I squirmed out of the scratchy embraces as soon as I could and stood apart while we scrutinized each other. I wasn't much impressed with what I saw but I was too polite to say so. Unfortunately, they did not match my childish courtesy. They looked at my slight body, my straight, light brown hair, my small, suntanned face with large hazel eyes. I must have seemed all one color and clearly the total effect was not to their liking.

"So *this* is the little girl," one said to the other, mournfully. "What a pity, she doesn't look a thing like her mother!"

All my life I had heard comments about my mother's beauty; I accepted them casually, complacently, the way she did. When she took her hair down, I liked to run my fingers through its incredible thickness and admire its dark gold color; it smelled nice, too. Everything about her smelled nice. I took for granted that it was all part of being a mother and putting cold cream on your face and scent behind your ears. And when I got to be that old—twenty-nine, going on thirty-five, I had heard my father say—I naturally expected to have milk-white skin, long blonde hair and blue eyes, too. But the visitors were planting doubts.

"Do you think she looks like her father?" the one who sniffled continued.

This I considered real madness. I would have to grow quite tall and straight, cut off my hair and wear trousers. I was relieved when the short one shook her head.

"Well, you can say what you like," she continued although the other one hadn't said a word, "but it must be a great disappointment . . . not of course that *she'd* admit it."

At that moment "she" came in and from the way she swept me into her arms I never would have guessed how disappointed she was.

Then the two ladies presented me with a small, fussily dressed doll they said they had brought me from Paris, and I was sent back to my own room. I pulled the black curly wig off the doll and stuffed her behind the radiator.

As time went by, it never occurred to me to doubt the words of the visitors. It became increasingly clear that my eyes were not going to turn blue or my hair amber-blonde and I marveled at the brave front my mother put up. I don't think I brooded about the subject or actively contemplated it . . . I hatched it. I must have sat upon the thought for years and kept it warm and alive. The sources of other revelations such as how babies are born were long forgotten but the memory of the callers and their revelation remained. My mother's enormous pride in my simplest accomplishments I interpreted as compensation, and her genuine interest in my appearance I took to be sympathy.

My mother was a poor dancer, I became an excellent one; she fought a lifetime, losing battle with her weight, I boasted

Occupation: Angel

about how much I could eat without putting on an ounce, and wore my belts unnaturally tight around my small waist. I shrugged off any suggestions about clothes or hair styles. I suppose it was my way of showing that I was not competing. I loved my mother deeply and yet the thought of her disappointment in my looks was an irritant—not constant but like an itch that came on only now and then by exposure to some unknown element.

After I was married and expecting my first baby, my mother died of pneumonia. My own childhood nurse, who was planning to come back and look after the new generation, came to call on me.

"I thought you might like to see this," she said, offering me a letter my mother had written to her some months before.

I gulped at the sight of the elegant but almost totally illegible handwriting that always spaced itself tidily until the very end when in a burst of afterthought it would circle round and round the pages and across the fold, filling in the marginal blanks with an elaborate labyrinth of calligraphy.

On the first page there was a reference to me: "I think pregnancy must agree with her, she walks gracefully and I thought her prettier than ever. . . ."

"Prettier?" I choked on the word. "Why did she use a word like that about me? And to you, who knew me so well?"

Pure bewilderment spread over Nanny's face. "Is it possible that you didn't know how much your mother loved your looks? Why, when you were a child, the household sometimes giggled behind her back; we thought it overextravagant the way she'd always refer to you as 'my beautiful little girl.'"

"And my mousy coloring?"

"*She* referred to it as 'subtle,'" Nanny said dryly, "and she liked it!"

Somehow, I didn't mind facing the fact that I'd been an idiot. As soon as I could go shopping again, I spent a delight-

Something Goes Whack

ful two hours picking out my first Paris hat. I don't know if it was becoming or not, but for the first time I felt pretty, inside.

* * *

And so, besides its unerringly right dramatic quality, *Lady in the Dark* had an especial appeal for me. Marshall Field liked it too and was ready to split the $50,000 investment with us.

I went around humming snatches of "My ship has sails . . ."; all was well in our best of all possible theatrical worlds. And then one evening my husband came home with a worried frown.

"It's the contract—the partnership paper for the production of *Lady in the Dark* is not right."

"Ah well, fix it up, dear," I said airily.

Our own, personal partnership had settled into a solid pattern. I read the plays and made the decisions. After the first investment and major loss, Howard had decided that that was my department. Howard read the contracts, wrote out the checks and took care of all of the business and financial problems. This suited both of us. I had no intention of stepping over into his department; if the contract didn't suit him, he should have it adjusted but to discuss it with *me* seemed a highly impractical approach.

"I'm not so sure that we can 'fix it up.' As it's drawn now, it links Marshall Field and me with Sam Harris as *un*limited partners."

"Limited, *un*limited . . . just so long as you've got your health," I said vaguely.

"I'm not worried about my *health*," Howard grinned, "It's my *wealth*—and, more important, Marshall Field's."

"How silly!" I exclaimed, "to start worrying about Marshall's wealth; why he has so much money. . . ."

But for once, playing Dulcy didn't work. I had to listen to

Occupation: Angel

the business details.

"An unlimited partnership would give us unlimited financial responsibility not only in this production but in the producer himself. Sam Harris is probably a completely honorable and trustworthy man; but it's too much of a chance to take." Howard didn't say it, but there was the implied danger of doing business with theatrical people. Visions of Florenz Ziegfeld who went into bankruptcy and escaped his irate creditors by blithely waving goodbye to them from the deck of his own yacht must have danced in his head.

"Can't you get him to change it?"

Howard shook his head. "He seems to be as stubborn about it as we are."

"Then compromise, just this once," I pleaded, "and do it *his* way. I do so love the play."

"I couldn't, even if I wanted to," Howard explained. "As you know, we invest through the family business of Cullman Brothers, and our by-laws do not permit such a possible financial liability. It's a bit complicated to understand, but under this agreement, even if one of the three partners were to *die* the other two would be responsible for any and all debts and liabilities."

There was an air of finality in his tone.

"What will Moss Hart think?" I wailed, clutching at a straw.

"He understands completely why we cannot go in and there is no shortage of available funds for financing."

Opening night, we went as the guests of Moss Hart; he always made me feel part of the venture. And as soon as contracts were standardized on a more realistic, limited partnership basis, we were part of practically all of Moss Hart's shows that followed. But even then, Moss made us feel part of that show . . . in everything but the financial returns, which were astronomical! Oddly enough, that didn't matter as money

Something Goes Whack

from our theater investments was rolling in at a great rate. Howard rarely questioned my decisions: I had lived up to my original boast of how often I was right, although often I must have been right for the wrong reasons.

In choosing *Charley's Aunt*, I wasn't exactly making a discovery! It originally had opened in London in 1892 and run steadily for 1,466 performances. It had opened in New York in 1893 and it has been said that, ever since, there has hardly been a month when the show has not played somewhere. At one time there were forty-four companies playing *Charley's Aunt* simultaneously throughout the world. It had been translated into twenty different languages, including Esperanto. There was no doubt about how the show played; the question was whether forty-seven years later a New York audience would find it ridiculously dated or charming as a period piece. I read it for the first time, felt a click inside and cast my vote for the latter; but mostly, it was a matter of putting blind faith in the skill of Joshua Logan as a director and José Ferrer's mad talent for nonsense.

The same was true for Maurice Evans' production of *Macbeth* with Judith Anderson. Knowledgeable people in the theater told Evans that the timing was not right; the public was restless and just wanted to be amused. But we agreed with Evans that, if the production were good enough, there would always be an audience. Both shows were smash hits.

One old-time producer said of us: "They're not just lucky. When you can make money even out of Shakespeare it's downright spooky!"

Old Acquaintance with Peggy Wood and Jane Cowl was a solid success and was followed within a month by *Arsenic and Old Lace*. Howard Lindsay and Russel Crouse produced it; they sent me the play to read one act at a time, thoroughly enjoying both my impatience and my wild enthusiasm. Since *Life With Father* (which was still playing to capacity busi-

Occupation: Angel

ness), they considered us "lucky money" and we stepped up from investing 20 per cent to 25. No one was more optimistic than I; and if the others mixed a little wholesome fear with their enthusiasm it was because they were smarter in the ways of the theater. In my ignorance, I did not know that on the stage murder was not supposed to be funny, and that insanity, unless treated in the most respectful, Ophelia-like fashion, was strictly taboo. For the others, it took real courage to defy every convention. I was too uninformed to have any qualms; to me it was a hilariously funny, preposterous, homicidal spree.

And so, I was not in the least surprised that all the critics and the theater-going public, not only in New York, but all over the world, saw it in the same light.

Arsenic and Old Lace was capitalized for $35,000, which was considered high in those days; we put up one quarter of the money for which we would receive one-eighth of the profits. Eventually, it enjoyed the fourth longest run on Broadway (*Life With Father* was destined to achieve the number one spot), lasted for two years in London, and broke records in Sweden and Argentina. I had an interesting and remunerative course in geography as the profits rolled in from Santiago, Chile; Lima, Peru; Melbourne and Sydney, Australia; Stockholm and Göteborg, Sweden; Rosario and La Plata, Uruguay; Honolulu, Hawaii; Buenos Aires and London. The producers were riding high and they had a real romp out of the business dealings; they varied the monotony of the regular statements by sending them out in the language of the country currently producing. Bewildered backers, unable to follow the language, wondered whether their statement said the play had just lost 1,000 kroner or whether that was their share of the profits . . . and if so, was that a lot of money? The producers gave a party for the backers and the backers gave another party for the producers; celebrations went on like a three-day Balkan wedding feast. We knew that we would make a lot of money

Something Goes Whack

on the venture, but we did not realize then that the play finally would gross between $8,000,000 and $9,000,000.

What with the truthful returns of sometimes as much as fifty to one that we were getting on our theater investments, and the natural tendency of all theater people to exaggerate, several highly apocryphal legends were being told about us. Friends and even slight acquaintances were crediting us with boundless sagacity and asking for the success formula. Any show of personal doubt or confusion was interpreted as, "Oh! So you won't tell, eh?"

At last, we began to be a little curious ourselves.

"Do you realize," Howard asked me, "that in a little over four years we've built up an equity in the theater which probably will be worth more than half a million dollars?"

"Is that a theatrical figure," I asked cautiously, "or for real?"

"For real!" Howard confirmed. "On my own financial statement. And, do you know something? I really got interested in the whole thing because I thought it would be a sort of hobby for us . . . something to do together that might also pay off."

Howard was in one of his rare, reflective moods. Perhaps it was because we were on a train going to California for a vacation and he was more or less trapped into physical inactivity.

"I wonder *why* we've done so well," Howard mused.

"You're an awfully good businessman," I suggested as one reason. Howard acknowledged this, modestly but with obvious relish. "And," I continued, "we have a nice relationship with the producers. They know we never want special deals; we have no boy friend or girl friend or relative looking for a part; we never try to impose our own opinions. . . ."

"In fact," Howard interrupted, "we never even *offer* an opinion unless it's asked for!"

We went along for quite a while, happily listing our own virtues. The train was hurtling toward the sunset where shaggy clouds still lined with sunlight were piling in the sky. Mostly

Occupation: Angel

the land was bleak; occasionally I'd glimpse a group of trees purple with distance against the horizon. Sometimes a town suddenly would erect itself, thrusting its ugliest buildings and factories forward. The train would hurl itself at them like a fierce missile, then miss and curve away again.

I thought we had come to the end of our virtues as angels; I pulled out two plays and wondered whether to start on the one in the blue or the red cover. Howard stared at them as if seeing a manuscript for the first time.

"There's our answer!" he said. "When we're all through figuring out the reasons . . . it comes down to selecting the plays! 'Right nine times out of ten,' " he said reflectively.

I felt a lump of honesty, like a touch of indigestion. "Oh, I don't know. I think when I first said that I was boasting."

"But it's true," he maintained. "You've done it! You can't figure just the plays we've actually invested in, you've got to count the total number of plays you've read and decided yes or no. Don't you see, if you turn down a play and it's produced and fails, that really counts one right for you. And on that basis you've come out right at *least* nine times out of ten! Where did you get the experience, the knowhow?"

Howard did not wait for an answer. The urge to ruminate was hard upon him and I was practically an innocent bystander. "It couldn't have been the time you spent on the theater magazine because you started out even there being right."

"Maybe it's just luck," I suggested mildly.

"Luck is a now and then thing; nothing that works nine times out of ten comes without some knowledge, some preparation."

"I know," I said, dutifully quoting Howard's favorite aphorism. " 'Fortune favors the prepared mind.' "

"Well that's right! And do you know something I've noticed about you?"

Something Goes Whack

"What?" I asked.

"You'll spend six weeks thinking about whether or not we should get a station wagon or a roadster. You've tried the piano in four different places; every piece of furniture in the house has been moved two or three times. You've never repapered a room or selected new chintz without consulting several friends and every member of the family: do I *really* like the pattern and am I *sure* I like blue in that room or would brown and white look better. But when you read a play, you come to an immediate, snap decision. You never dither around; you never ask anyone's opinion, in fact you don't seem to *care* what anyone else thinks."

"I don't mean to be arbitrary about the theater," I protested, "I didn't think you *wanted* to read the plays and decide."

"I don't," Howard said. "I just want to know why you're so tentative with one and so independently sure with the other."

"Oh, that's easy." I was relieved. "I guess it's because in decorating I don't see the finished product as a whole. I have to grope and feel my way. But with a play, I see it as a whole, and I can visualize every bit of action, the entrances and exits as it goes along." I was explaining it as much to myself as I was to Howard. "And then if it's right, really right, somewhere around the end of the third act something goes 'whack' and I know we're in!"

"Something goes 'whack'?" he repeated.

"Yes. I think I have a kind of built-in divining rod. You know, the kind people carry around when they're searching for water. . . . Well, I have it, not for water but for plays."

Howard was not terribly well attuned to what he called my pixie side. And I wasn't at all sure that we were on safe ground, but I continued.

"I'm not even conscious of this thing until I've read a good play and then I can feel it, as if a divining rod hit the ground.

Occupation: Angel

If I don't like the play, I don't feel *anything*. Once in a while, and that's the confusing part, I don't feel a real whack, just a little quiver. Then I stop and brood about the play and wonder if the construction is really sound and who they're planning to cast in the lead, and all that sort of thing. That's when I'm likely to think too much and go wrong."

I felt that I had made a sensible and complete disclosure.

"So . . . a responsible businessman like myself has been writing out checks for up to $25,000 a play because you thought you felt something go 'whack.' Well, I'll be damned!"

I must have looked hurt because Howard said hastily, "I'm not questioning your judgment. In fact, I've just been telling you how good it is. I just don't think that you come to decisions the way you *think* you do. No doubt, you had more specialized training in drama than you remember. How about college—what did you major in?"

"I majored in panic," I said to myself, and aloud I said, "I only stayed in college half a year—there wasn't time for much specialized study."

"How about earlier . . . much, much earlier. Didn't you meet a lot of theatrical people?"

"No," I said, "that early, family brush with Bohemia was short-lived. My mother didn't really like theatrical people. And for the next few years my contact with the theater was confined to an occasional, little-girl, Saturday matinee."

I paused, trying to recall that period in my life, and then memory came flooding back. I knew that my mother went to see something called *The Poor Little Rich Girl* and she had been so affected by it that she rushed home in the middle to see that I was not, like the child in the play, being neglected or ill-treated. She came right into the nursery, kissed me and woke me up. And I remembered hearing her talk about a play she liked. It was called *Pygmalion* and somehow I gained the impression that the star of the play was eminently respectable because she was called by her married name: Mrs. Patrick Campbell.

We saw *Seven Keys to Baldpate, Daddy Long Legs* and a jumbled series of musicals. We were allowed to see anything with George M. Cohan because Father said that anything he played in was absolutely "clean." Father was equally vehement about the *Evening Journal*; he said that he would not have it in the house because it was a "dirty" paper. I interpreted both of these statements quite literally and associated them with Father's preoccupation with germs. I did experiment by surreptitiously rubbing my white kid gloves over several newspapers at a stand and, to my surprise, the *Journal* didn't come off any dirtier than the others. But I hadn't been able to figure out how you tested to see if a play was clean.

Howard was obviously convinced that I had a past he didn't know about. And the more I thought about it, the more I realized he might be right. I hadn't told him about some things because I hadn't thought of them myself for a long while. The hard times, one might call them, although they held their own curious delights.

4

Flashback....

Old summers with almost forgotten sensations came curving in. . . .

It was strange, how quickly I seemed to feel beneath my feet the soft, aromatic pine needles of the woods; or ran with the wild freedom that only a city bred child can truly appreciate, along an endless, deserted beach. Usually, there was a slightly briny smell on a fresh breeze; but sometimes I'd come upon the half-decayed carcass of a fish and then the smell would stab my nostrils with the sharpness of pain. It was at the odd, infrequented hours and especially when I was alone that I felt I took the beach unaware. The rough sea was a bully and a braggart, heaving and dashing itself wildly up the beach. When I felt I'd had enough of it, I'd cup my hands

Flashback. . . .

and call out to it: "All right, all right. Uncle!" I liked it better when it tried to fool you by appearing quite calm and then convulsed and powerfully hurled a wave up on the shore with a great sigh. Often it would leave behind long, slimy strings of seaweed which obviously belonged in the sea and would stretch after it, taut and straight like fingers straining to grasp it. Most of all, I loved the skittish little waves that always seemed to be dropping curtsies and leaving brief souvenirs of fans edged in beige lace foam. This was the bonus for which the sandpipers waited, their long gray legs twinkling under motionless little round bodies and their eager eyes spotting invisible goodies.

There were fat, delicious days of July when we'd be pulled shivering from the cold water and told to snuggle down on the hot sand to bake ourselves warm: innocents indulging in the ancient cure of arenation.

On cloudy days we might dig for worms: plump, juicy ones that were bigger than the minnows that we fed rather than caught in the inlet.

Then there were the hot, wide, white glittering days of August when the sun was rude and it was safer to lie in a hammock on the side porch, reading books you loved or ones whose chief charm lay in the fact that everyone said they were too old for you.

Either because there is a natural affinity between sand and children or because the community was predominantly Catholic, there was a surprisingly convenient number of contemporaries. At that age we rarely sought out kindred souls but rather adhered to the unwritten rule of choosing as playmates the boys and girls next door or across the street who came within two years of the same age, and then adjusting our personalities as best we could to suit.

Then came war, the "war in Europe," as we called it. Our family was unable to adhere to the ostrich policy which

Occupation: Angel

seemed to occupy our neighbors because my father was in the champagne business and the vineyards in Rheims were in danger of being completely destroyed. A reassuring touch came from my elder brother who confided in us that he had it straight from the father of a friend that, mainly, the war was a publicity stunt to promote Pathé News and make people go to the moving pictures.

Abruptly, the United States declared war and everyone was intensely united and violently suspicious of anyone with a German-sounding name or accent. Mother even contemplated switching hairdressers until he solved the problem of changing his name from Schmitt to Smith. We saved tin foil for some obscure reason and emptied our banks to buy War Saving Stamps. Everyone learned the words to "Over There" and sang it as if it were the "Star Spangled Banner."

We begged our parents not to use German as their secret language. They couldn't use pig latin, as we were already highly proficient in that and could snap out a warning to each other: "Ix-nay, ook-cay is-ay ooking-lay," before pinching a sweet from the kitchen. They tried French but neither of them spoke enough really to communicate. That was how we caught on to why Father was always going to Washington—he was doing something for the Secret Service but we couldn't put up a flag with a star. On the whole, we found it an exciting and highly diverting time.

I began to take more interest in the stickers that almost overnight appeared in myriads of windows; they had one, two and sometimes three stars. Then a sprinkling of gold stars began showing up. Speeches were made about how proud those families with the gold stars should feel but when I saw people coming in and out of the houses they did not look at all proud; they looked sad or vacant-eyed. And I felt cold when I passed the houses. Gold stars, silver stars . . . the night sky no longer seemed happy to me.

Flashback. . . .

The Armistice was a fine thing to celebrate but it did not settle our own family problems: many of the vineyards had been destroyed and Father had lost a great deal of money. It was not too serious, of course, but it would take a few years for him to feel back on his feet. Meanwhile there was a good deal of conversation about economy measures, but I didn't notice any appreciable change in our way of living. There was talk, too, of Prohibition. One day it was an absurd, unbelievable bugaboo: then overnight it was a certainty and my father's business was completely wiped out.

My family never did things by halves; where formerly we had wallowed in comfort and ease, we now wallowed in uncertainty and, finally, poorness. First the car and chauffeur went; then the house in the country, the servants, and even a few pictures and valuable pieces of furniture. I constantly had the sensation of putting my hands out to touch something I knew, only to find it gone.

It was the beginning of the bad time. We moved uptown to a cheap apartment and that seemed to be the final severance from most of my city playmates—the country ones were already gone.

My contemporaries were slipping, with varying degrees of speed, into adolescence, but I lagged behind. I was in a new and frightening sphere, I yearned back to the happier years and somehow my body seemed to pay homage to my heart. Some of the girls took to carrying powder compacts and making a great fuss about it being the wrong day for them to enter the swimming meet. I had no such problems: my body remained flat, unripe, and I never failed to show up for swimming practice. This was somehow misinterpreted as personal valor and I found myself captain of the team. But swimming was extra on my tuition and it soon became expedient to drop it. Everything cost more, except reading. I was desperately lonely and unhappy. The days had such a frightening same-

Occupation: Angel

ness about them that the passage of time was hard to judge.

It was Phyllis, a former classmate, who opened up a road of escape. She telephoned and asked me if I wanted to make a date for a certain Saturday afternoon; we'd go downtown to lunch and see a matinee, Dutch treat. I was cautious and wanted to know how much it would cost. She said we could do it for under two dollars each and I decided to spend some of my birthday money.

I was impressed with Phyllis, who seemed to know her way around. She took me to The English Tea Room on West 48th Street where we splurged on a fifty-five-cent lunch and then walked down five blocks to Gray's bargain basement, cut rate ticket office. Unsold theater tickets were disposed of here and anything might show up. There was a posted list of available shows with the prices of the tickets but these usually ran too high. The trick was to wait until the very last minute, and keep an open mind as to what you wanted to see. The procedure at Gray's ran something like an auction sale; someone stood on a raised platform, a telephone next to him; a call would come in and he would shout: *"Six Characters in Search of an Author . . . orchestra, ninety cents!"* We would dash forward, thrusting our dollar bills into his hand and receive a slip of paper, convertible to tickets at the box office. Then we sprinted over to the theater, often arriving breathless just as the curtain went up. Maybe that's why even now I cannot see a curtain go up without a small gasp.

An uncle supplied me with an allowance of $1.00 a week and by the most careful budgeting I could manage two and, once in a while, three excursions a month. There was never any question about varying our routine: Phyllis and I always went to the Tea Room which was a sort of Green Room of twenty theaters rolled into one. The walls were hung with old playbills; the establishment was run by two elderly, stage-struck but efficient sisters. They knew how to cook and loved

Flashback. . . .

good food; they also loved actors who in turn loved their food, so the circle was complete. If a young actor could afford it, he ate there and if he couldn't afford it he almost *had* to eat there; no place else was as generous and patient about credit. It was the place to go for sympathy if your show flopped; it was the place to go for congratulations if it succeeded.

The telephone was attached to the open wall—news traveled fast. The Tea Room was warm, friendly and intimate: it was the Café des Deux Magots of Broadway, with a respectable New England touch thrown in.

For Phyllis and me there was no such thing as a bad play; some just were better than others. At first we went to anything we could get for our price, then I began to look for outside opinions. I started to read Heywood Broun, Alexander Woollcott and Percy Hammond and this opened up a whole new aspect of pleasure in the theater.

Everything about the theater fascinated Phyllis but she had an especial affection for the actors—male or female; it was the kind of devotion the British have, where they will see their favorite stars in any show. I was play-struck and for me the actors were alive only on stage; I had not the slightest concern with what they ate, wore or did in their private lives and I had no loyalty to offer them. Far from coming between Phyllis and me, this separation of interest served only to unite us. It was like finding out that your roommate really *preferred* the white meat of chicken. And so, we divided the plays in fair and equal parts: Phyllis could worship the cast while I took the playwright.

Phyllis and I were at different schools; we saw each other only on those occasional Saturday afternoons and after the exhilaration wore off, I'd be depressed by the bleakness of the following two or three weeks. Our family life was in chaos: I was filled with an ever accelerating sense of disaster. I saw no possibility that things would get better and so I could only

Occupation: Angel

presume they would be worse.

Financially, my father was floundering helplessly; anxiety had brought on a recurrence of an old heart ailment and my already troubled sleep was sometimes punctuated by a sharp cry of pain and terror when he suffered a heart attack.

I was afraid to stay in school and be an additional drain on the budget; I was even more afraid to leave and be forever bound to some dreary clerical position. Where could I go without an education?

I tried to compromise and searched the want ads for a summer job. The business division of the New York Telephone Company had just such an opening for high school students over sixteen. I was fifteen but I crossed my fingers and lied about my age. My looks were against me: I had a childish face and my clothes were even more incriminating for they were the well-cut, sturdy models left over from our more affluent days. However, I was at least a year ahead of myself in school and probably that is what turned the tide in my favor.

I got the job—such as it was. It gave me twenty dollars a week base pay and extra for overtime work; it also gave me long hours of almost unbelievable boredom.

"Filing and research" had sounded intriguing, but I found that the filing consisted of flipping endlessly through trays of alphabetically arranged cards and seeing that Mrs. J. Herrington McSniff came *before* Mrs. Peabody T. McSouse. And the research was nothing more intriguing than finding out if a customer was excessively delinquent (the service turned off but the instrument remained) or deceased (the phone removed).

Each morning I took the subway at 145th Street and Broadway and rode down to the Wall Street section. I had to change at 42nd Street, follow a green line on the ceiling to the shuttle and then leave the shuttle for a third subway train. It was new

Flashback. . . .

to me, an all too charted but fearsome and unfamiliar course. In the evening I reversed the course, and each trip was a battle either for a seat or for a spot where I was not intimately flattened between strange bodies.

For days at a time my conversations at the Telephone Company were limited to "Good morning," "Thank you," "Oh, not at all," or "Yes, these are completed."

I felt I had nothing in common with the hordes of other girls who worked in the department. A few, like myself, were summer replacements, but most held permanent jobs which they must have taken out of necessity and kept out of inertia.

* * *

When I got back to school in the fall, I winnowed out my experiences and the unpleasant parts fell away like chaff. I was the only one who had worked and most of my classmates envied me.

I must have played my role of Tom Sawyer in his fence-painting scene very well because the following summer another friend, Gertrude, yearned to apply for a job *with* me. Her family, who were very well-to-do, were unenthusiastic. But Gertrude was adamant and romantic: "We are the wave of the future. We are two characters straight out of George Bernard Shaw and I wouldn't miss this chance for the world."

Gertrude eventually talked her family into letting her try for a job. This time I scanned the ads and chose carefully. The Metropolitan Life Insurance Company seemed a likely prospect and the location on 20th Street and Madison Avenue would simplify our transportation problems.

We went down together in answer to the ad and we filled out endless papers. I decided to hold onto the extra year I had given myself and listed my age as seventeen. We seemed to be passing all the requirements nicely and were sent into another room. There we were told that we were to have a physical examination. I expected to show my teeth and stick out my tongue but we were put into small, separate cubicles like bath houses and told to undress and drape ourselves in a sheet. We were horrified but too timid to protest. We felt strangely insecure and vulnerable in our togas. Gertrude whispered to me through our partition: "Do you think this is . . . er, you know, all *right?*"

"They couldn't be planning anything *wrong,* on such a major scale!" I whispered back and we drew courage from the contact.

A brisk woman doctor checked us one by one. It was the first time I had been examined by anyone but the old family doctor who had taken care of me since infancy. I was surprised to see that she poked, prodded and tapped her right knuckles on two left fingers over chest and lungs in the familiar manner. Still and all, it seemed somehow shocking to have a *woman* do it. Gertrude and I concluded that she too was a wave of the future—a middle-aged one.

We passed the physical but before we could be placed in the right niche or have our salaries established (the ad had offered $20 to $25 a week) there was another hurdle . . . an I.Q. test.

That was my day! Every bead, ball, triangle, square and round peg fell into place as if by magic. I drew straight lines between given points and traced routes like a qualified surveyor; I rapidly underscored "yeses" and "noes" to a long series of questions; and tied up scrambled words with their definitions. I felt wildly self-confident.

Gertrude joined me on a long bench where we sat awaiting the results of our tests and she confided that she had suffered from a sort of stage fright and might well have failed—and if so, after all the fuss, *what* would she tell her family! We did not have long to wait.

I was very pleased when the personnel aide congratulated

Flashback. . . .

me and told me that I had come out with an exceptionally high intelligence quotient and that consequently I would receive $24 a week and would be in Exploration-and-Alphabetical-Classifying.

With infinite relief, Gertrude learned that she had passed. She would be in Inter-office Communications. Her job was a sort of high-class messenger girl.

We started work the next day. Gertrude was in a trance of delight. Her conversation from then on was freely peppered with such remarks as: "The vice-president made an interesting remark to me this morning. . . ." Upon careful probing her honesty and humor would prevail and she would admit that he was probably the seventh, *assistant* vice-president and that what he said was, "Thank you, very much. Please shut the door on your way out." But it had seemed mighty fascinating at the time.

Gertrude had been obliged to make a small compromise with her family. The chauffeur had to deliver her in the family car in the morning and pick her up every evening. This was an embarrassment for her but it was one that I was more than willing to share. With a little persuasion the chauffeur saw our problem and agreed to drop us and meet us around the corner, letting us travel one block on foot. In the first flush of enthusiasm for her business career, Gertrude made me solemnly promise that I would stick to the job for the whole summer. I assured her that I had every intention of doing so but to make it more binding she insisted that if either one of us gave up, the quitter would be obliged to pay the forfeit of a piece of lingerie. (We were so recently out of Ferris Waists that we found the very word "lingerie" eminently satisfactory.)

My actual work was no more stimulating than it had been at the Telephone Company; indeed, if you substituted the words "policyholder" for "telephone subscriber" it was almost identical. And alas, exploration and alphabetical classifying

Occupation: Angel

turned out to be synonyms for research and filing!

The girls with whom we worked were somewhat different from the ones I had met the previous summer; most of them were high-school graduates and some of them had taken business courses. There seemed to be very little financial pressure at home and their salaries mostly went to buy new dresses or linen for their hope chests. Gertrude and I did not form any friendships, but we had each other and the neighborhood proved more interesting to investigate than the crowded, narrow and sunless streets of the lower financial district. Sometimes we would spend our lunch hour exploring the 14th Street district with its wild hodgepodge of European and Oriental stores. It was a vast improvement on the summer before; and I did not feel so alone. Even after the first of August when Gertrude sent me an embroidered *crèpe de chine* slip and went to Lake Placid for the rest of the summer, I was content, if not fascinated, with my job, and I saved like a miser. The money would help to pay for my schooling.

* * *

I had been back at school a few months and was sorely missing my weekly salary when another chance to earn money came my way. A friend of the family had been offered a part-time job as a reader for Goldwyn. The work consisted of reading the new books as they were published, writing a short review, a full synopsis and a brief commentary on the possibilities of the book for a movie. For this she would receive $8.00 a book (and it made no difference whether it was a simple claptrap thriller of two hundred pages or an involved intellectual tome of twice the length by Virginia Woolf). There were private reasons why our friend did not want to turn down the job; on the other hand, she confided, it took her almost a week to read *any* book, and then she couldn't always remember the plot, to say nothing of writing about it.

Flashback. . . .

I couldn't help saying that I wished I had the chance. "You do . . . ?"

And so, I read the first book and wrote the synopsis. Our friend turned it in under her name and we held our breaths. It was accepted and she was given two more books to do. I never knew why she felt the need to pretend that she was writing the reports, but she generously insisted upon giving me the full pay. I found that besides attending school and doing my homework, I could turn out three reviews a week. I was making as much money working nights and weekends as I had full time over the summer.

In a short time I was working under my own name. The friend had found the deception onerous and after a few weeks had told the head of the reading department that she had an offer of a better job but that she had a friend who was most anxious to take over.

"As a matter of fact," she had added in a burst of combined candor and dishonesty, "*she* wrote the last three for me because I was too busy."

The head of the department checked and said that *my* work was equally good; in fact, I wrote in much the same style. The work was mailed to me and, fortunately, the question of age never came up. I suppose he figured that a friend was a contemporary.

But my days of glory did not last long. Goldwyn was undergoing a complete reorganization and merging with Metro; temporarily, the reading department would shut down. I was obliged, for the first time, to go down to the offices to get the money due me and sign some sort of release or receipt. I went down one Friday afternoon after school. A newly arrived executive for the West Coast office took one look at me and all hell broke loose.

I caught broken snatches of "teenagers . . . child labor" and "What were you running here, a children's camp?"

Occupation: Angel

I escaped with my check and the realization that, even if they reorganized and reopened promptly, I was not going to be among those present.

I took stock of my situation and realized that, although I would not officially graduate from the convent until June, I had enough credits to get me into Columbia at midterm. A basic knowledge of typing seemed indicated and I decided to throw in shorthand for good measure. By this time I was used to doing, or at least trying to do, two things at once so I signed up for my regular courses at Columbia in the mornings and enrolled in a course in typing and shorthand for the afternoons. For some reason I had the conviction that the stenography course would be a breeze. I was humiliated to discover that all those pothooks, shaded lines and careful alignments were almost beyond me.

I had achieved the status of a college student which somehow was terribly important to me, but I was only barely making the grade. I was becoming more acutely conscious of our financial problem and it constantly intruded when I tried to concentrate.

Mother took a job in a dress shop and she made it sound like the greatest possible fun. But she didn't succeed in fooling me; I had played that game myself. If she faced moments of fatigue or even despair she never showed it. Years of having her clothes made at the best couturiers had given her solid fashion training and the simplest dress that she selected for herself immediately took on an air of authority and chic. No outsider would have guessed how close we stood to poverty. Perhaps that's why no one helped. And there are no charts or guides for the inept, newly poor.

By spring, my father's health was, if anything, worse; the heart attacks were more frequent. He had another worry: his good friend the playwright Aaron Hoffman had developed a heart condition also.

Flashback. . . .

"I guess we're both getting to be a couple of old dodoes," he said. But from the way he smoothed down his already tidy black hair and lifted himself by his chin until he looked two inches taller, I knew he did not mean it.

"Fifty-two isn't so *terribly* old," I said soothingly.

"Thanks." He sounded uncomforted.

When Aaron Hoffman had recovered sufficiently from a sharp heart attack, he went down to Atlantic City. Father also went down to keep him company for a week. I wished that I could have gone, too. I enjoyed the company of Mr. Hoffman whom I knew not as a glamorous, rich and successful playwright but as a gentle, rather shy man, who was able, however, to establish a relationship between us which was immediate, personal and familiar, no matter how long the interval between visits. When he talked to me about books and plays there was no touch of condescension. He made me feel older, and apparently I made him feel younger—an odd but mutually enjoyable alchemy of time.

The Sunday afternoon Father returned from Atlantic City, I was in bed with the grippe. Father wasn't much of a family man: he had a cup of tea with Mother; stirred up a little excitement around the house; complained that my two older brothers not only had been wearing his best remaining ties but were slobs; thanked God for the fact that their feet were already too big for his shoes, and whirled out again. But before he left, he spent a few minutes chatting with me and rubbing my back—a special treat.

"Mr. Hoffman sent you his love . . . he's very fond of you."

That night when he came home we were all asleep. Shortly afterward there was the now familiar sharp outcry in the dark; but this time fever stood between me and consciousness like a thick fog. Then I heard my mother sending a brother out into the pre-dawn to get help—get a doctor and *quickly!*

It was the thin brother, now broad-shouldered and enor-

Occupation: Angel

mously tall. He came back, clutching a reluctant doctor as a fierce terrier might bring in a rat. There was another cry of pain and fear. My brother pushed the doctor into the room ahead of him, then followed and closed the door.

I stood alone, outside the closed door. I was frightened and helpless. In a little while the doctor emerged by himself and looked at me.

"Daughter?"

I nodded.

He flipped open his notebook: "What was your father's full name?"

I answered mechanically, slowly letting the past tense sink in.

"Middle name?"

"He had no middle name. . . ." and then as if this seeming deprivation proved too much, I burst into sobs.

* * *

Most of the callers that day came to see my mother. Aaron Hoffman gave me the feeling that he was there for me. I was still running a fever so he sat on the edge of my bed and absentmindedly pulled my hair. It was his way of petting me. After a moment or two he got right to business. He knew that I was studying at Columbia and that I was also taking a course in typing and stenography.

"As you know, your father spent the last week with me—oddly enough we talked a good deal about you—and I said I wished you were mine! Maybe it's thwarted paternalism—maybe I'm just interested in molding a young mind. In any case, we made plans for your education and your future. . . ."

I shook my head. "It won't be possible. Father didn't leave *anything*."

"He left you a perfectly good brain," Aaron Hoffman said firmly, "let's start using it. We'll step up the plans a bit. In

Flashback. . . .

another month or six weeks your classes will be over; I ought to have my health back and be ready to start on a new play. I'm giving my secretary the summer off—and that's where *you* come in! You'll work, and you'll learn. In the fall you'll go back to college full-time but holidays and summers you'll work with me. The School of Journalism may help; but *I'll* teach you the craft of writing! And someday . . ."

At this point I think we both went off into separate daydreams. I daresay he was waving aside the apologies of the head of the English department at Columbia and agreeing to outline the new and improved Hoffman method.

I was older, perhaps a well-preserved twenty-two, and in response to the cries of "Authors! Authors!" I tenderly led out onto the stage my aging, probably fifty-five-year-old, co-author, Aaron Hoffman!

He knew the mentality of the young. And he knew that a growing dream could, at least temporarily, push aside sorrow. It was a good crutch for the following weeks but then something went awry with the plans. His health did not come back and within a month I was back at the same funeral parlor, feeling as if I had lost my father twice.

5

Push Aside Sorrow

This time the numb, lifeless feeling did not last as long. But, for a brief time, I felt unable to focus on my life or look for work.

Fortunately, work came looking for me. Andy Rice, who had known my father, offered me a temporary job replacing his male secretary, Jones, who was off on leave. It was not a secretarial job in the usual sense and for the temporary, summer span, even my shorthand and typing wouldn't be needed.

Mr. Rice was playwright, lyricist, gag man and general punster all in one. He was high in a now extinct profession: a vaudeville-act writer. It was a strange field, one in which he received good money—when the actors remembered to pay him—but unless he wrote a skit for the Scandals or the Follies, he got no billing or glory. Andy Rice was the wit-without-credit behind the Mort Sahls, the Shelley Bermans and the Danny Thomases of the day. But other actors usually knew the source of the material and he was much sought after. Most of the acts were written to order for a particular actor or team; it was customary for the actors to pay a lump sum for this tailored-to-order material and a percentage of their earnings as long as they continued to use the act.

In those days popular vaudevillians would be booked at private parties, clubs and the regular vaudeville circuits all over the country, usually winding up in a burst of glory at their own particular heaven: the Palace. Sometimes, Mr. Rice would have free time and then he would write a playlet or two which he would file away, waiting for the day when someone called for help in a hurry. He ran an open shop: Theater Material For Sale. And it would be my job to run the shop, a one-room office on Broadway in the Forties while Mr. Rice stayed at home and wrote. I would see anyone who came in, make appointments, sort the mail, answer the telephone and protect the man at work from disturbance.

Unfortunately, there was no one from whom to protect him. It was summer and the season was slack. A whole week went by and no one came in.

There was a filing cabinet crammed with rumpled carbon copies of patter—the jokes they used between soft shoe dances and songs—skits, blackouts for reviews, and playlets. I smoothed them out and typed fresh copies of everything, but it was the playlets that interested me most. I'd read the descriptions of the sets and the stage directions over and over again to myself. I finally figured out for myself that "stage left" was the audience's right-hand side; "up stage" and "down stage" were more ambiguous. Sometimes I'd read quickly, just to get the story line, then I'd go back and read the dialogue aloud in the empty office to see if I could judge

Occupation: Angel

how long it would run; it was always under half an hour.

That left little time for subtleties or character development, and so the characters were delineated by easily recognizable symbols. All wives were virtuous and the "other woman" wore a long, tight black dress with ropes of pearls to indicate *her* status. All young girls wore breton hats on the back of their heads as if the baring of the brow automatically invested them with youth and freshness. Clean-cut young men wore stiff straw hats but occasionally so did a slicker. However, if he was a real scoundrel, he was obliged to wear a mustache and if he was a seducer he would appear, regardless of the time of day, in a flowered damask dressing robe. This cleared the deck for action and notified the audience whom to watch and what to expect. I recalled early novelists who, even with ample time to develop their characters, preferred to alert their readers in a similar manner by giving the characters appropriate names like Michael Strongheart or Patience Worthiness.

It was rather like taking a course in the drama with only the playlets themselves to serve as my teachers.

I tried to guess which of the miniature plays had met with the most success and then looked them up in the records, which I found were not very well kept. More for my own curiosity than from a sense of office efficiency, I tried to straighten out the apparently hopeless tangle of who had what material and what payments had been made on it. The latter was the easiest part; apparently, Rice was known as a softie and almost no one paid promptly, if at all.

Today, a popular television writer sees his material burned up in twenty minutes but in those days similar material, if the act clicked, would be good for years and should have paid regular dividends.

Mr. Rice finally stopped by one day to inquire rather wistfully if there were any royalty checks. There weren't. I suggested that I could write some dunning letters. He was a sympathetic man and he looked pained at the suggestion. Finally he agreed to let me send out some.

"But not to this one," he said, going over the list of delinquents, "his wife just had another baby. . . . And not to *this* one; the poor old girl is beginning to slip, having a bad time of it."

I wound up writing less than a dozen letters. A few came back stamped "address unknown," which renewed in me the sense that actors were not real people but chimeras subject to unaccountable disappearance. But there were two responses with promises to pay "soon" and three actually were shocked into enclosing payments on back royalties amounting to a couple of hundred dollars. I was considerably cheered but in my eyes all actors were automatically suspect.

Each week I conscientiously read *Variety*, painstakingly working out the peculiar verbal shorthand in which it was written. I was well prepared in advance to understand the headline for the crash of '29: WALL STREET LAYS AN EGG!

Andy Rice dropped in at the office, looking for checks that were seldom there. One day he was feeling offended: Lou Holtz was opening that night in a musical comedy and was supposed to be using a lengthy series of lyrics which Rice had written for him. But Holtz denied using the material on the road tryout and was not paying for it. I suggested that Rice go to the opening that night, catch Holtz and demand payment. Mr. Rice shook his head; everybody knew him and, even if he did not order the tickets in his own name, he'd scarcely be past the box office before Lou Holtz would be alerted and could switch material. The whole idea was distasteful to him. A writer didn't have much protection and clearly here was a man who did not intend to fight for his rights.

It often was customary to precede an evening opening with

Occupation: Angel

a matinee—it served as a kind of paid rehearsal.

"How about letting *me* go, right now, to the matinee?"

"Do you know the lyrics?"

I pointed to some of my handiwork, the new, neatly typed original and two carbons of the work in question. "Besides," I boasted, "I don't have to depend upon memory, I'll take it down in shorthand!"

"I just hope they don't turn you away from the box office as a minor," Rice said, giving me three dollars for my ticket.

No one had ever turned me down, and I reminded him that I was seventeen.

"Well, you don't look it," he said glumly.

I had a good seat in the second row, I was armed with a pad and three sharpened pencils. I figured that no first-class spy could do more.

Lou Holtz appeared in a couple of scenes, but he did not use any of Andy Rice's material. I began to worry: maybe he sensed something. Or maybe he really *wasn't* using the lyrics, which would be something of a financial blow.

Then Holtz came out again, casual, carefree and self-confident. He strummed a guitar and began to half-sing, half-recite the lyrics. I flipped open my pad and jotted them down as fast as I could. Even though the words were familiar to me, I had trouble getting all the pothooks right side up and shaded correctly. I completely forgot about everything else and concentrated on my work.

I could feel prickles on the back of my neck; I was conscious of being stared at. I glanced up and felt like a prisoner trying for a night escape and finding a half dozen beams of light converging on him. Holtz had seen me busy scribbling and had strolled over until he was directly in front of me and was twisting his head around making a mock effort to see what I was doing. The people around me laughed and stared with him. And so, a perfect showman, seeing that he had a bit of

Push Aside Sorrow

extra amusement, Holtz extemporized and threw in something about most people just laughing at his jokes but there was a girl copying them down so that if she didn't catch on the first time around, she could laugh when she reread them. He was quick and quite funny but I felt that I might drown in the wave of laughter that followed.

When intermission came, I couldn't decide whether I would be less conspicuous if I remained huddled in my seat or if I went out into the lobby and tried to lose myself in the crowd. I didn't have to make a decision. The head usher came down and pinned me to my seat with a baleful glare. Mr. Holtz, he told me, was quite disturbed by my writing down what he was saying. I could get into real trouble doing that. And didn't I know that it was against the law to steal other people's material? I wanted to say that that was just what Mr. Holtz was doing—taking someone's material without paying—but the truth of the matter was I was pretty well scared. It was my first attempt at playing Mata Hari and I hadn't learned the rules. I wondered if I should tear out the sheets and try to swallow them.

When the usher addressed me as "little girl," I had a clue. I shed a year and a half as a starter for sympathy and said that I was writing a thesis for the drama class at my school.

The usher looked somewhat mollified, but still suspicious.

"Oh, yeah? Well, what are you writing down every word he says for?"

"That's the whole point of the thesis. . . ." I was beginning to warm up to the story I had created. In my mind it was not so much a lie as a game. "Half the class is writing on tragedy, half on comedy. And in the comedy group a lot of the girls are using Charlie Chaplin as an example, but *I* have chosen Mr. Holtz because I think he's much funnier."

"You do-o-o-o-?"

"Much funnier," I affirmed. "And I'm trying to prove that

it is not necessarily what he says but how he says it that counts."

The usher saw no need to listen further.

"Okay! I'll tell Mr. Holtz what you said, and how you're writing a whatchamacallit."

"Thesis," I supplied softly.

The usher returned in a few minutes and was quite agreeable. "Mr. Holtz was real pleased—after he understood what you were doing. He likes young people to take an interest in the theater. And he thinks it was real smart of you to catch on to the words not counting so much as what he *does* with them. He does a reprise of the song in the second act and he said he'll take it nice and slow, so you can get it all down."

I thanked him and he started to leave when he remembered something.

"Oh yeah, and when you finish that thingamabob he says you can send him a copy. Maybe he'll autograph it or somep'n."

Mr. Holtz got a copy of what I was doing much sooner than he expected. In fact, he got it the very next day; all signed, sworn to and attested.

Andy Rice reported all the details to me; he seemed terribly amused.

"You should have heard what he said about you—you and your *thesis!* He nearly chewed up the scenery."

I took a prim stand. "I think he's dreadful, trying to cheat you! Why, he's *dishonest!*" I said, conveniently forgetting my own elaborate lie.

"Oh, I don't know," Andy Rice countered, "maybe he'd have paid . . . eventually. He's really a pretty nice guy . . . and a great performer! You see, you can't judge actors by the usual standards. They're better—and they're worse—than most people. Anyhow, different."

I wasn't sure whether he was confirming or negating my

grandmother's point of view. But on one point they surely agreed: actors were different.

The episode created a bond of friendship and now when Rice would say, "You're a funny kid," I knew his words held a touch of approbation.

It was late summer, and the season for catching the acts in their tryouts. New acts were booked into second-rate vaudeville houses in places like Astoria, Newark or New Haven before they were ready for the big time. This was a sort of trial run, the time to see if the skit was playing smoothly, if the timing was right. It corresponded to a regular play's period of rehearsal and out-of-town tryout. Usually, the author of the skit would "catch" (one never said "see" in this case) the act in time to make notes, study, polish and rewrite wherever necessary. He was not only author but also served as a consultant producer-director.

"I thought Jones, my male secretary, would be back by now. He was invaluable on these trips," Rice said gloomily. He was preparing to "catch" an act in Long Island.

"Couldn't I help? I can take shorthand too, even in the dark," I reminded him. "And I'd love to do it."

He brightened. "Maybe you could, just this time, because we could come back the same night."

I was delighted with the victory, but still anxious to vie with Jones in usefulness.

"How about the other trips? Did Jones stay overnight with you?"

He nodded.

"Well, how about me? I'm a good traveler."

Andy Rice shook his head. "No! Impossible!" he said firmly.

I thought it over for a minute and then the solution came to me. Obviously, it was too expensive for me to have a separate room. The other way the two of them could share one. But when I tentatively asked if that was the problem, Mr.

Occupation: Angel

Rice seemed quite cross. I thought he didn't like the idea of my knowing that business wasn't very good.

"Not at all," he snapped. "Of course we had our own separate rooms. But at my age—fifty—I don't intend to be seen trailing around with a seventeen-year-old little girl."

He never had sounded so snappish or strict with me before, so I stopped the discussion there. But I puzzled for a long time over how illogical he was. There he was planning to take me across the street to have dinner with him at the Astor Grill, and he didn't seem to mind being seen with me at the theater. Why would he be ashamed of me in front of a hotel clerk . . . or was it out-of-town people? It was confusing.

After dinner we took the subway out to Astoria. We did not have to hurry because the skit was one of the features of the bill and therefore would be the next-to-the-last act to go on.

Once again, I was fortified with a shorthand pad, a clutch of pencils and a boundless sense of excitement.

We slid into our seats during a softshoe routine, two acts in advance of the one we had come to see. I scarcely could contain my excitement and I decided then that never, never, even if I had had the talent for it, could I be an actress. The tension and the breathlessness were too much.

When the curtain went up on the playlet we had come to see, I mentally stood on tiptoe; I was ready for magic. But the act started poorly and I felt a sense of panic. Thirty minutes wasn't a very long time in which to establish characters, develop a plot, catch the imagination of the audience and make them laugh. In fact, it seemed an almost impossibly brief time but that same half hour could prove an interminable period to sit through if the act wasn't going well.

I sat, pencil poised, waiting for the special words from the author that would be the key to setting everything right, but for the first few minutes all I heard from him were a few grunts of disappointment.

Push Aside Sorrow

At last there was an appreciable ripple of laughter; this was followed almost immediately by a bit I thought unexpected, fresh and really funny, but it fell away on the heels of the first laughter.

"Spacing!" Mr. Rice said crisply.

"Spacing!" I wrote in shorthand: a little ball on a light stick leaning backward, then another little ball swinging into a heavy curved line like a hammock. The economy of effort was fascinating and satisfactory to behold. Then I realized that it stood naked and alone on the page. How would I ever know to what it referred? In parenthesis, I scribbled the next line of dialogue so that I could place the comment. I was beginning, in a groping fashion, to catch on.

For the next twenty-five minutes I transcribed abrupt, laconic comments and the author was as impersonal as if he were hearing the lines for the first time. As I could not anticipate the remarks, I continued to identify the place by the line of dialogue that followed. By the time the skit was half over, I had become bold enough to make separate comments of my own on the page above: "Audience did not respond here—did they miss the *point* of the telephone conversation?" I began to wonder why some lines which I had considered very entertaining when I read the manuscript brought forth only a mild snicker while others caused a real guffaw. I tried using question marks; then I hit upon the idea of working out a somewhat inaccurate laugh meter and I indicated the heartiness with from one to three stars.

Toward the end there was a really hilarious denouement; the laughter carried over and muffled another few minutes of rather obvious explanation.

"Cut!" the author ordered.

And I did not need to identify the place. We both knew where the new finale would come.

Going home on the subway, we explored the notes I had

made and Rice patiently explained the reason for each contemplated change. Perhaps he was using me as a sounding board for his ideas.

"What's that, on the page above?" he asked.

I explained my own notes and he was delighted with my laugh meter.

"You were right, noting only one star where the wife opens the box of flowers. . . . I was worried about that; I'm going to cut it out entirely."

"Oh, no!" I objected. "When I read it, that was the part I liked the best. And even if this audience didn't laugh, I think it's funny, and endearing!"

Andy Rice shook his head emphatically. He was adamant but clearly not angry. I suppose no author ever minds the protest that his writing is sacred and too good to be cut.

"No," he said, "it has to go. You see, you're only partly right; it *reads* well, but it doesn't *play* well. That's the hardest thing to learn but it's the final, true test for the theater . . . the moment of truth. Remember that!"

In the weeks that followed, we were a familiar sight—the ill-assorted couple going from one scrubby little break-in house to another and always taking the train or the subway home the same night.

After one of the skits had made the grade and been promoted to a big-time house on Broadway, occasionally we'd go again, just for the fun of it. Then we'd nudge each other and recall how one bit or another had laid an egg in New Haven and had required careful work to bring it up to its present sparkle.

I had pleasure out of the rest of the acts but, instead of finding them surprising and therefore more interesting, I missed the sense of intimacy in knowing what the next line would be. I found myself wondering how it had played for the first time in a break-in house.

Push Aside Sorrow

If you really were in the know, you arrived late enough to miss the opening act and left before the last one; these were the spots traditionally reserved for the fillers-in.

I still loved the legitimate theater best, but I had no intimate connection with it and that made a difference. Besides, vaudeville in those days was far from all cheap boffo and slick comedy. It was a rich upstart in the theatrical family and both producers and owners of legitimate theaters eyed it with envy.

The vaudeville managers faced almost none of the usual expenses of a legitimate show. All they supplied was a theater, an orchestra with a glib leader who could act as a fall-guy or a semi-master of ceremonies when needed, a box office, a few stagehands and some ushers. The house was never kept dark for a month, waiting for a particular show to come in; and there was no heartbreaking failure if the show flopped.

The actors came already equipped with build-in material, costumes, music and sometimes their own scenery. Even a poor act was no great loss—it was only one tenth of the whole and, in any case, next week there'd be a complete new show.

Vaudeville had plenty of money to spend and had persuaded some of the great names to cooperate. They were still boasting of the fact that Sarah Bernhardt had given a condensed version of *Camille* and that the renowned David Belasco had produced a *Madame Butterfly* especially for the Orpheum Circuit. Between legitimate shows, Ethel Barrymore did a thriving business with a stewed-down adaptation of James Barrie's *Twelve Pound Look*.

That season of 1925, between the assorted vaudeville shows and the big, new movie houses that were also offering a stage show, one might have seen a singer, hoofer and impersonator called Milton Berle; another singer named Morton Downey; a fiddling comedian named Ben K. Benny who changed his name to Jack Benny so that he wouldn't be confused with Ben Bernie. And Bernie, incidentally, had an impertinent,

Occupation: Angel

young piano player in his orchestra named Oscar Levant. Mary Hay's dancing partner was Clifton Webb; Walter Pidgeon was a popular baritone and George Raft was billed at the Rialto Theater as "the fastest Charleston dancer." Charles Correll and Freeman Gosden played the piano and the ukulele in an act called *Harmony Syncopation*—they were getting secondary billing. The following year they switched to a comedy act and called themselves *Sam 'n' Henry*; who can say whether it was the eventual change of name or the precise training they received in the vaudeville school of five-a-day that eventually made them the hits of the radio world as *Amos 'n' Andy*.

Gertrude Lawrence was being wooed with an offer of $3,500 a week; eventually, she succumbed and played a full season. But Mae Murray, who was offered a contract for $5,000 a week, turned it down because she was making, in those tax-free days, $7,500 in the movies.

It was an era of big money but the awards went to the big-time stars and not much rubbed off on the little writers who supplied their material.

The fall theatrical season was well under way and I had almost forgotten about Jones, whom I was replacing, when a letter came saying that his mother finally had passed away and he would like to return to his job if it was still open to him.

I said goodbye to Andy Rice.

"Don't leave. I'll find enough work for both of you," he promised.

There really wasn't enough work for two. Besides, I had looked over the books and the poor fellow wasn't up to being burdened with two semi-indigent, half-orphans of assorted sexes and ages. I was aware that the job originally had been offered to me out of kindness of heart; I also was aware of the fact that I had more than earned my salt. The confidence Andy Rice had placed in me was my greatest reward: and I

Push Aside Sorrow

knew as little of self-doubt as I did of insecurity. I felt very grown-up, we were revising roles and I was reasoning with my mentor.

"Business *isn't* very good," he conceded, "but it'll pick up after the first of the year. . . . Are you sure you can get another job? Maybe you'd better come back next week and we can talk it over?"

"I'll be working," I said with the full confidence of someone who doesn't know what she's talking about.

6

My Education

Oddly enough, I was right.

I looked over my situation thoughtfully, trying to see it from all four sides. I was seventeen and couldn't get away with padding my age more than a year. I had had half a year of college and had held down three temporary jobs. The first two had given me nothing but a familiarity with the intricacies of the subway system and a wholehearted distaste for mass employment of any kind. The third job I had enjoyed thoroughly and I felt that I was good at it but few, if any, serious dramatists were able to afford a girl Friday and I was instinctively wary of theatrical producers or managers. The girls who worked in their offices had something I did not have: a savvy, an awareness, a wise sophistication which I was barely percep-

My Education

tive enough to realize I lacked. I decided against trying for anything further in the theater.

I was a good typist but my shorthand was shaky—fine for personal notations but likely to collapse under steady dictation; however, I had no intention of becoming a stenographer or a business secretary anyhow. I had no illusions that I could earn a living writing but I wanted somehow to be associated with the literary world. There was still the field of books, newspapers, magazines or communications of any kind.

Aside from courage and a boundless enthusiasm, I had frighteningly little to offer. Fortunately, I did not think of that.

The Sunday want ads yielded two possibilities which attracted me, and with a fearful confidence I pondered *which* ad to answer.

The first was for an intelligent, executive young woman to be in charge of a bookstore specializing in the drama and rare books with fine bindings. It promised unlimited opportunities for the right person. I was charmed with the general ambience, but the "executive" touch gave me a slight twinge of doubt. Perhaps that is what spurred me on to answer the second ad and write the McNaught Newspaper Syndicate. I no longer recall just what *they* asked for in the ad; all I know is they got *me!*

I had answers to both of my letters. The first response was from the man who owned the bookstore; he was particularly interested in the fact that I had run an office for a writer and we set up a meeting. He was almost speechless upon finding that his "executive and intelligent young woman" was a seventeen-year-old who somewhat unsuccessfully claimed to be eighteen. I mistook his bewilderment for indecision and tried to clinch the deal by assuring him that I was well pleased with the prospects of the job. Gently, he broke the news that I was too young. Just as patiently, I tried to point out to him that

Occupation: Angel

we must take a long-range view and that my youth was a highly temporary liability. The poor man must have been utterly confused by my quiet certainty. We continued to chat, purely in the abstract, about books and plays and writers; I thought we were getting along very well.

"You'd better leave now," he finally said, "before I make a complete fool of myself and hire you!" And then he added: "Perhaps, if I don't find the right person, I'll call you and talk it over again in a week or two."

"By then," I said, as much in sorrow for him as for myself, "I probably shall be working for a newspaper syndicate. That's the *other* ad I answered," I explained.

His face looked absolutely blank down the middle and incredulous around the edges. He gave me a silent, weak wave goodbye.

Two weeks later he telephoned me: "I've talked it over with my wife and she agrees with me that probably it's crazy but, as I can't find the combination of qualities I'm looking for, I might try you out as my personal assistant—and then at the end of a year or two . . ."

I told him I already was working for a newspaper syndicate and that I liked it very much.

"The *other ad?*"

Even on the telephone, I could feel the italics in his voice.

"Well, good luck to you," he said when I had confirmed his query. "And I can't decide whether the syndicate should be warned or congratulated!"

It is just possible that the office manager dreamed up the title of editorial assistant to please me—actually, I was just a step above an office boy. But they had an office boy, and the exact location of my seat below the salt was of small importance. I remembered my older brother's Alger books; all I wanted was a start. The heroes of the Alger books always had aspired to the top. So I looked the syndicate over carefully.

My Education

The president not only was uninspiring but his work mainly was confined to the business aspects; the vice-president was the traveling salesman of the firm and if he hadn't sold cartoons and comic strips, articles and features he would have been just as skillful with a line of tractors and trucks.

I had no ambition to either title. It was the two editors whom I was supposed to assist who fired my imagination. I would be content, I decided, when I got to be an associate editor. Somehow I must have felt that the title would automatically invest me with the warmth, the wit and the wisdom they both had in such abundance. They passed on the work of aspiring syndicate writers, negotiated for the serial rights of books, dreamed up new features to be assigned, found the writers and edited the work before we sent it out. Apparently, this did not keep them too busy as each carried on private, additional work of his own.

Charlie Driscoll, a quiet, bespectacled, scholarly man, wrote blood-curdling, moderately successful books about pirates; having been born and raised in Kansas, he was an authority on sea lore.

Fred Knowles was the best self-educated man I have ever known and his side line was a couple of highly remunerative public relations accounts. It used to amuse him to show me his own tricks of the trade. He got remarkable coverage, especially in small town papers, on any material he sent out. He used to say that it was due to a simple formula of establishing an honor system: don't be a pig and always give good value for what you want. When du Pont wanted to promote buttons, he made contacts in Paris and sent out an excellent fashion column with a Paris dateline accompanied by ready-to-print mats of suits and dresses featuring buttons. All of it was free. He never for one moment pretended that he did not have an axe to grind; he just made it an attractive proposition for the fellow who owned the sharpener. If he wanted to pro-

Occupation: Angel

mote Pillsbury flour he would send out, under his pseudonym, Betty Crocker, a first-class article on cooking, winding up with a couple of recipes and, if the recipes happened to call for a specific brand of flour, nobody quibbled.

As if the editors' dual occupations were not enough, they shortly added a third: my education. By now, I'm not sure just why they took it on. Charlie Driscoll could not spare as much time, he had a good, big, Irish brood of his own to look after but Fred Knowles was childless and had both an urge and a great talent for teaching. I suppose the real proof of it is that I never thought of him as "teaching"—he was just sharing his knowledge and exchanging views. Actually, what he was doing was experimenting with his own theory of related study and in this he probably was a quarter of a century ahead of the standard system of education. In my case, it was impractical to attempt languages and he automatically bypassed science and mathematics but he taught, or at least exposed me to, art, history, philosophy, literature and the drama as unified subjects. He might start with a casual discussion of a couple of Italian artists which would lead us eventually into the study of the entire Renaissance. And either he or Charlie Driscoll would just happen to have a couple of books in their bottom drawer which they would be glad to lend me. They taught me how to use the city's museums; the public library was just around the corner; so was the Metropolitan Opera House and it didn't cost much to buy standing room. They encouraged my interest in the theater and if there was a really fine revival, especially of Shakespeare, it was surprising how often one or the other just happened to have a Saturday matinee single ticket that he wasn't going to use. Sometimes, one of them would drop a question and let me try to figure out the solution on my own. Sometimes it took me weeks or even months.

One day I listened to them discuss Shakespeare's *Henry VIII*.

My Education

"You've got to admit it's a pretty poor play," Fred Knowles said.

I wasn't in the habit of doubting him but this shocked me.

"When a man writes thirty-eight plays, averaging two a year, he's entitled to come up with a comparative turkey now and then. As a matter of fact," he continued reflectively, "I have a theory about *why* that play is weak, and it has nothing to do with whether or not Fletcher wrote it, or collaborated. See if you can figure it out. Your work with Andy Rice should give you the clue."

First I read the play. We were in an Elizabethan period then but I had been slipping up a bit on my homework. I was aware only that the cadence of this play did not please me as much as some of Shakespeare's other works. I suggested that, as it was one of his last works, maybe he had gotten too old; but I was reminded crisply that a man in his forties was not exactly in his dotage. It was difficult to vision how it would play. I remembered Mr. Rice's admonition that it wasn't how a play read but how it played that counted. I suggested this to Fred Knowles, who said that possibly I was on the right track. I skipped lunch a couple of times and spent the hour in the library. There I unearthed the story of Shakespeare's part ownership in the Globe and the Blackfriar's Theaters and the account of the great fire in which the Globe burned to the ground in the middle of a performance. I sat musing about this for a few minutes and then the words on the page seemed magnified: *During the first performance of Henry VIII!* I felt like a detective who, against all odds, has brought in the criminal.

"Did it have anything to do with the fire at the Globe? And is it possible that Shakespeare never did see his own play in production so that he could cut and polish and make changes?" I was breathless with excitement.

Fred Knowles' face lit up with approval and he stopped

Occupation: Angel

biting on the end of his pipe long enough to agree: "That's *my* theory, anyhow."

When I talked to some of my friends at college, I was aghast at how dull their studies were by comparison. And later, when I went back to Columbia to take extension courses, I found that my professors lacked the wit and capacity to stimulate that I had grown used to.

One of my duties at the syndicate was to edit and then type on sheets for mimeographing a couple of unimportant daily and weekly features. Sometimes there would be a few misspelled words over which I would cluck happily; a good, fat grammatical error made the day for me. My responsibilities were slight but I rarely confined myself to those.

Once, at the very beginning, I rushed in too fast and nearly lost my job. At that time the syndicate had started to send out a daily bit of homespun philosophy by Will Rogers. It was based on the news of the day and was short, pungent, deliberately illiterate. We sent it out by wire. Most of the papers carried it in a box on the front page and paid accordingly. The day Charlie Driscoll left the office with an abscessed tooth, he failed to remember the Rogers release. Fred Knowles had left word that he might not be back at the office that afternoon. Happily, I proceeded to act as editor and I translated Rogers' salty, cowboy idiom into prissy English, utterly emasculating the copy. Then I started to send out the telegrams. I had carried the torch of learning on to the next town, or at least as far as Trenton, N.J., when Fred Knowles came back, saved the release from slaughter, the syndicate from a lawsuit and me from losing my job!

I learned a little humility; and the fact that the incident was treated as a joke was perhaps the most humiliating part of all.

There must have been times when my editors doubted the comfort of having an assistant. I was always looking for more

My Education

work to do; it was the only apple I could bring my teachers. And so, little by little, they gave me more assignments.

For a while I was assigned to a young woman who did a combined fashion and society feature. We referred to it as the "Who Wore What, Where" series. She went to the races, the openings and the smart restaurants, observed the people and later sketched them and described their clothes. She followed the smart set from North to South, West to East. I was to serve as guide and aide during her stay in New York. Although I was a complete novice, I thought I could help in memorizing the costumes so that she could sketch them accurately later on. However, one outfit had a way of blending with another until I was no longer positive which was the full-length and which the seven-eighths coat. I was appalled at how much escaped me but I eventually caught the trick of concentration; and I learned to take a mental snapshot of the woman as a whole and then list the details in my mind.

Shortly after that, Ted Coy, the famous Yale football hero and husband of the ill-fated Jean Eagles, was signed up to do a series of six weekly articles on football. Over a hundred newspapers were convinced that this was just what their reading public wanted.

This series was to be my special responsibility: weekly copy was to be in on Tuesdays; then it had to be edited and meet the Friday deadline. The editors didn't bother to tell Mr. Coy that his work was to be handled by anyone so far down the line.

Perhaps one of the editors had the gift of intuition, or perhaps he had seen Mr. Coy's capacity for Martinis at luncheon; in any case he took the precaution of seeing that I had a backlog of two finished articles before we started.

I was delighted with the number of corrections I was obliged to make. But the job of editing was nothing, it was getting the material in that counted. In no time Mr. Coy had slipped

Occupation: Angel

a week, and then was two behind. That meant that I no longer had the cushion of the extra articles. I worried, and I pursued him relentlessly by telephone. Sometimes he wouldn't bother to answer the phone at all; sometimes the blurred tones, the failure to identify me or what I wanted and even the mumbled apologies presaged further delay. The poor man was taking his marital problems hard and was bolstering himself up the only way he knew how.

In a sober and contrite moment, he telephoned the editor. Driscoll's face crinkled up in amusement. He motioned to me to listen in and then, complaining that there was a buzz on the wire, asked Coy to repeat what he had said.

"I merely said," Coy reiterated, "to please tell that poor old dame who pursues me not to worry, I'll get everything in on time from now on. I suppose I should tell her myself, but I have a horror of old-maid newspaperwomen!"

But the following week he was off again and the final article was even later than any of the others had been. I could not locate Mr. Coy; he had not shown up at home for two days and a check of his club and his favorite speakeasy failed to turn up a clue. I had a couple of hours in which to get out the piece or notify a hundred papers to change their schedules.

Carefully, I reread the first five articles, underlining bits that seemed especially good to me. Then I wrote a sort of résumé of Coy's own football wisdom, added a few observations of my own and padded the whole thing with a batch of current, Ivy League statistics.

This time I proceeded cautiously. I presented our problem to the editor and added: "Perhaps we could send *this* out."

He glanced over the sixth and final article which bore Ted Coy's by-line. "But of course we can send this out! It's fine. When did he write it?"

"It's a forgery!" I admitted. "I didn't know what else to do but could I be arrested or something for writing it and sign-

My Education

ing his name to it?"

"Send it out! And let's just not discuss it with the police at all."

For days I walked around on a cloud: I had written something that was syndicated in over a hundred newspapers—and I was a forger! I wore both thoughts snugly wrapped around me with equal delight.

I had a third mentor for the times when I wore out the first two. He was Rube Goldberg, who used one of the syndicate offices as his studio. In the corner of the room there was a washbasin and he made the error of inviting me to use it any time I chose. I chose quite regularly. It is doubtful if there was a girl with a cleaner pair of hands in all the Times Square district. Sometimes he'd stop to talk to me, sometimes I'd look over his shoulder and watch the Tuesday Ladies' Club come to life, or a Goldberg Invention develop. That was in the days when a cartoon was a complete entity in itself. You looked, laughed and enjoyed it without being obliged to see the three previous ones to know what it was all about.

One winter day, I watched Rube Goldberg finish a cartoon about the day in spring when men officially broke out in straw hats. He was planning a trip and was getting ahead in his work.

"Isn't it hard to keep thinking of so many funny situations?" And then I added in embarrassment, "Or does everybody ask you that?"

"Yes—to both questions. Mostly," he reflected, "it's a matter of some instinct, and a lot of training. You learn to look for humor, systematically approaching the subject from various angles. For example: I may start chronologically with the first of May. And I'll say what's funny about May Day; May poles; dancing on the green; Bolsheviks; spring in general; love; proposals; engagements; weddings? The more sub-divisions you can work out, the more clues for jokes. Get it?"

Occupation: Angel

I said I thought I did.

"And as for the instinct for humor, it can be developed, to a degree, but it has to be there in the first place. It's one thing to see the humor yourself; it's quite another to be able to channel it into your own method of expression and pass it on to others. Do you follow me?"

"Not completely," I admitted.

"Put the soap down! You make me nervous twiddling with it that way. You don't *have* to wash your hands to come in here, y'know."

"Yes, I mean, no . . . thank you very much."

"All right! When things are funny, you have to see them in your own particular area. Now, as I am a cartoonist, I tend to see all situations translated into my own medium. If something amusing occurs during the day and you want to recall it later in your own mind do you see it, or hear it?"

Now I was on familiar ground: "Both, I think. It's like a blackout sketch in a revue, if it's short; and like a full play, if it's long."

"That's good," he said, encouragingly. "And, of course, a simple situation may become funnier, based upon how you personally interpret it, react or become involved in the incident."

I nodded happily. "That's like the girl I saw in a stationery store last evening; she bought some valentines . . . and I thought it was funny."

"You've lost *me!*" he said.

Rapidly, I sketched in the background for him: the very, very pretty girl fluttering over a large collection of valentines in the group labeled *To My Sweetheart—Male*. She was terribly earnest in her quest and read each card carefully before rejecting it. I found myself following her search; it was important to get just the right one. I thought about how strange and how nice it must be to be in love, or even to have a very

My Education

special beau for whom one took such care.

I had received my package and my change but I lingered a few minutes, anxious over the girl's selection, involved in her romance.

She turned down two more cards and settled on a pretty one that said simply and directly: *Because I Love You.* Shyly, she held the card up to the salesman and her voice was soft as she said: "I'll take half a dozen of these, please."

"Write it," Rube Goldberg ordered, "and submit it to a magazine."

And so a few days later I stayed after hours and typed my story. Fred Knowles suggested a small change which sharpened the incident and we decided to submit it to the old *Life* Magazine, the American counterpart of *Puck*. One of the syndicate's cartoonists had dropped in and watched us curiously.

"Let me see it," he said. He read it quickly and nodded. "Tell you what," he said, tucking the story into his pocket, "I'm doing some free-lance cartoons for a new magazine that's starting—I'm on my way over there now and I think I can place this for you!"

But I was reluctant to speculate with what might be a fly-by-night venture.

"I think you're foolish," the cartoonist said and looked to Knowles for support.

Fred Knowles shrugged. "She has to make up her own mind. After all, she isn't exactly a novice. . . ."

I looked up in surprise, barely in time to catch a slight wink.

". . . as just last week she wrote an article that was syndicated in 103 newspapers!"

I returned the wink.

A few weeks later, the cartoonist dropped by again. He had an advance copy of the new magazine and showed me his cartoons in it. I admired them.

"I wish you'd had something in it too," he said generously.

Occupation: Angel

I pulled out my top drawer and proudly displayed an acceptance slip and a check for $25 from *Life*. It was the first and last check I ever received from them but at the moment my future seemed pleasantly resolved.

I accepted the cartoonist's congratulations. Then I looked back thoughtfully at the magazine's cover with its highly stylized Beau Brummell equipped with high stock, shiny top hat, curled sideburns and a monocle through which he haughtily observed an unlikely pink butterfly.

"*The New Yorker*," I read aloud, doubtfully. "Well, I don't think *that'll* be around very long."

Prophecy was not one of my talents.

* * *

In those days, only the more enlightened offices conducted a five-and-a-half-day working week. I felt very fortunate to have Saturday afternoons off but, aside from an occasional matinee, this free time represented little to me. My job was the most interesting thing in my life; I had practically no social life, neither dates with boys my own age nor exchanges of girlish confidences. In part this explains why I understood only the most elementary aspects of the facts of life. Sex was a subject Mother did not care to discuss. She had given me a dandy little bird-and-bee book when I was ten years old and had made it plain that volume two concerning humans would come as a wedding present from my husband.

What I did know, I knew in the classical sense; even the phrasing was formal and technical as I had gathered most of my information either from the Bible or from the *Encyclopaedia Britannica*. But I felt confidently well informed.

In preparation for an emergency which seemed unlikely to arise in my life—the gay bachelor who would invite me to a speakeasy—my three syndicate guardians decided they'd better initiate me themselves. I was allowed one drink only, and that

My Education

was straight. They taught me to sniff and taste warily and then to toss the drink down like a veteran newspaperman. This seemed to amuse them enormously.

One Saturday, the four of us dropped into a new speakeasy. There was an attractive-looking young woman sitting alone at the bar. I thought she must be waiting for someone who was late. She observed us carefully for a few moments and then, with a broad and easy smile, she came over and said: "Only one girl to go round seems like short rations, wouldn't you like me to join you?"

This seemed to me a surprising but very friendly gesture and I was shocked when my friends abruptly rejected her overtures and the bartender told her to scram.

I suggested that the girl was just being friendly and one of my escorts said: "She was 'friendly' all right—that's her business." Then he added: "She's a doxy . . . a lady of the evening."

"Oh!" I said loudly in final recognition. "You mean a whore!" Only I pronounced it "w-a-r."

"A what?"

"A w-a-r," I repeated distinctly. "You know, 'Thou art the w-a-r of Babylon.'"

Charlie Driscoll caught on first.

"She means a 'whore'," he howled.

For once Rube Goldberg was beyond laughter.

"Where did you ever hear such an expression?" he wanted to know.

Charlie Driscoll's merriment had beclouded his glasses. He took them off and wiped them as he answered for me. "Obviously she never *heard* it, she *read* it—in the time-honored method of all precocious youth."

The word "youth" at once alarmed the bartender.

"Hey! What is this?" he wanted to know. "Where did *you* meet these men? And how old are you? I don't serve no liquor

Occupation: Angel

to minors!"

Somebody hastily thrust a bill at the bartender in settlement of our check. He seemed completely unnerved by the whole incident and, as we left, was saying aloud to no one: "Christ Almighty! Whores and minors! I wisht I was back in business in a respectable corner saloon! I swear I do!"

I wasn't invited out again for a long time.

7

Search for Identity

A couple of years slipped by; it would have been easy and comfortable for me to stay put, but I knew the time had come for me to make a break. I needed to make more money and to move ahead faster. I had the sense of "Time's winged chariot drawing near" even though, at twenty years of age, it traveled at a snail's pace.

I sampled several jobs: I read once again for a moving picture company but the work was somewhat spasmodic and therefore impractical; jobs with the good publishing houses were so sought after by the college crowd that they paid a mere pin-money salary; the small literary agency I worked for proved interesting but also unremunerative.

Publicity turned out to be the answer. It not only paid well

Occupation: Angel

but it was exciting. Each job presented a whole new picture, it was a challenge and took every bit of imagination and ingenuity that I had to offer. I started a publicity firm with a partner whose energy matched my own. When she left to become Cholly Knickerbocker on the *Journal American*, what seemed to be my loss turned out to be a bonus. She was besieged by ambitious matrons who wanted to give a ball or a country fair primarily for diversion and secondarily for the benefit of anything handy. When these ladies asked for advice, my friend gave it. And as a result, they generally telephoned me for an appointment. I remembered my early instructions about giving full value and I tried to make it work both ways. The magazines and newspapers trusted me. If I had to make a choice, I preferred to lose a client rather than a newspaper contact.

Balls and benefits led to beach clubs in the summer and night clubs and restaurants in the winter; travel movies led to chambers of commerce. I was highly eclectic in my taste. I stopped talking about doing publicity and started to discuss permeating the consciousness of the public; instead of making a splash, I recommended a deep psychological penetration. I had the gold-lettered sign PUBLICITY removed from the door of my office and substituted a smaller, more discreet PUBLIC RELATIONS. The two words instead of one made everything twice as expensive; my clients didn't seem to mind.

Bonwit Teller had been bought by Floyd Odlum and a major change was contemplated. Mrs. Odlum was to be made president, the first woman president of a major store. It would be an important story which would require careful handling and they were looking around for a new head of public relations. Someone suggested me and I was interviewed. The job would carry with it a good deal of responsibility and prestige. It also meant full time work to the exclusion of everything else, and would mean giving up my own office. The offer was

Search for Identity

too good to turn down, and yet I was not entirely certain that I wanted it. Perhaps that's why I got it!

My new office on the executive floor of Bonwit Teller was comfortable, bright, well decorated and came equipped with a small, blonde secretary. It also boasted southern exposure and a fine view of the large terrace of the penthouse apartment opposite. Later, when I recounted the story of my romance and marriage, the southern exposure, the view of the penthouse terrace, and the secretary figured prominently in the story, but at the time they were minor assets.

There were rumors that Hortense Odlum had been given the presidency of Bonwit Teller as a Christmas present, like a bauble that her husband would drop in the toe of her stocking. There were a lot of other rumors, too. I ignored all of them. I preferred not to know but to play it straight; this gave me a greater air of sincerity.

The initial part of the work was fascinating, and I was enchanted with the prospect of playing Pygmalion. Although Mrs. Odlum was hardly Galatea, she was pliable and gracious and she fed my Pygmalion pride by never going off on independent tangents of her own. Prior to a press conference, if I read the release I had written in her name and said: "This *is* how you feel about the role of women in business, isn't it?" I could depend upon her to say that this was *exactly* how she felt. She had a trick of answering a direct question in so slow and deliberate a manner that the most perfunctory comment sounded like a considered judgment. Sometimes I felt that I had created a false character, but sometimes I felt that sheer intuition had caused me to write the very words Mrs. Odlum had intended to say herself. All this left me personally in a pleasant state of uncertainty and absolved me from any sense of charlatanism. The story of a woman president made very good copy but one can go on creating an image, permeating the consciousness of the potential client and making a deep

Occupation: Angel

psychological impact on the public just so long. Then the story was played out and I was obliged to concentrate on the fashion aspects of the job.

These were the depression days; the crash was recalled as a period of relative prosperity. It seemed as if half the graduating class of every good finishing school bypassed college and headed straight for the employment bureau of Bonwit Teller. If we had been obliged to call an emergency meeting over a weekend, the social register would have been more useful than the telephone book. The debutante department was filled with debutantes who were supposed to attract other debutantes—and they did, only those attracted were also broke and they came to call or look for a job, not to buy.

The junior department was very proud to have a genuine princess on its roster. In fact the department set more store by the title of Princess than Dolly Obolensky did herself.

"Being a princess is not so much of an advantage," she confided to me one day. "I get only five dollars a week more than the other girls!"

At first, I was wary of a quick friendship with her. I was afraid that the title was seducing me, too. But princess or chimney sweep, it would have been impossible not to be beguiled by Dolly and her family. They were all handsome, intelligent, utterly without pretension, and they were living as best they could in a new world. I think they felt they conformed completely to a conventional pattern of behavior, although nothing could have been farther from the truth.

Many of Dolly's family stories would begin with: "Did I ever tell you about the time my Uncle Sergei . . . ?" Gradually, I deduced that this man was an uncle of her mother's back in Russia and had nothing to do with the handsome, debonair Serge Obolensky about town whom she also called "uncle."

"Then Serge Obolensky," I reasoned, "is your father's

Search for Identity

brother, of course."

"No," she said patiently, "you're getting all mixed up again —maybe you're thinking of Uncle Nicholas?"

I ignored the potential complications of Nicholas and stuck to the uncle I had in question. "But you call him 'uncle,' don't you?"

"Certainly, I call him 'uncle'! He's much older than I am." Dolly's logic was not always easy to follow.

"Then he isn't really an uncle at all; but he *is* related?"

"Of course he is related, otherwise why should I call him 'uncle'?"

(I had a feeling that there was an old vaudeville team that used to do a routine something like this.)

"We *always* call all older relatives 'uncle' . . ." Dolly explained, and then, taking my ignorance of Russian customs into consideration, she grinned and added: ". . . all *male* relatives! It is more polite, no?"

Dolly's English, marked with an accent, gave the impression that she was translating spontaneously from the Russian. She spoke French fluently and could get along reasonably well in three or four other languages, but she had not the slightest conception of mathematics. Her life was all like that, and she accepted the vagaries as one accepts uncertain spring weather. She was realistic but modest and she never confused her accident of birth with achievement.

One afternoon, about closing time, Dolly dropped into my office. She had had a rough day and missed lunch.

"Come on," I suggested, "let's have tea someplace in the neighborhood. It'll do us good."

But Dolly lived nearby and she suggested that we go home and rummage in the icebox.

We walked past the priceless ikon that hung above a shabby, overstuffed chair whose springs scratched the floor, and went into the kitchen. Dolly opened the icebox, which contained

Occupation: Angel

remarkably little. She pulled out a large, half-empty crock of caviar and spread it thickly on black bread. "There's nothing else to eat," she said dramatically. "Nothing! Not even a slice of ham. An empty icebox makes me feel so poor!"

The caviar eggs were large, gray, and unsalted, the most expensive kind. We both ate with a good appetite and it seemed as though I had never tasted anything so good. I thought of the usual cocktail-party bits of caviar that looked like lumpy black ink soaked into crackers.

"It's funny," I said, "but at the moment I feel very rich."

Dolly looked surprised, but pleased. It was the inability to offer a guest a choice that bothered her, she explained.

"Ah, I wish you had been able to come to the party we had last week! You should have come in late, after the fashion show you had to attend was over."

I wished that I had been able to come, too. The fashion show had proven a horror: two of the professional models had gotten drunk and I had been obliged to cope with them and finally substitute an amateur who was not only inadequate but the wrong size.

"It was *such* a good party . . . music, singing. . . ." Dolly's eyes grew radiant with remembered pleasure. "And at the end, there was a toast—and we all smashed our glasses against the fireplace, that is, where there should be a fireplace. It was lovely."

"But you *broke* all your glasses?" A touch of practicality came struggling to the surface of my mind.

"Oh, it didn't matter; by that time we didn't have any more wine left anyhow!"

* * *

One of the bonuses of my job was the need to follow the fashionable crowd. After I came back from a few weeks in

Search for Identity

Palm Beach, I went down to see my old editors at the McNaught Syndicate. We still would have an occasional lunch or drink together. They teased me about my expensive sunburn and my smart suit.

"Do you know something, I've just discovered that I hate the word 'smart' when it refers to fashion," I said.

Fred Knowles nibbled on the end of his pipe.

"And how well do you like fashion?" It was more of a thought tossed into the air than an actual question. I didn't think it necessary to answer him. I waited and answered it to myself, later.

When I was in Palm Beach, the Artists and Writers had been there on their annual outing. Some of them were old friends from the Syndicate days and it had given me a great sense of nostalgia for the warm, wonderful people I had known then. It was possible to find people you could love in any sphere of work or life, but it took more searching in the fashion business.

It occurred to me that the writers and the artists I had known were often moody and sometimes difficult, but in the balance they had more to offer. When they were struggling, you felt that the battle was within themselves but as they achieved success they grew from within so that they seemed to rise upward through their own silhouettes.

In the fashion world, progress was quite different. Most of the people rose in a fiercely competitive manner; sharp heels dug into the bodies over which they climbed. The victors seemed to be perpetually carrying a battle flag which must be planted just ahead, and ahead. There was a singleness of purpose which would have been admirable had the flag represented something like liberty or truth but, if one looked sharply, it only said "I." For the first time, I was encountering warfare in the pink jungle of the fashion world and I didn't

Occupation: Angel

like it.

Something about the constant battle stripped a woman of her femininity, layer by layer. And the more successful the woman, the louder the jangle of gold bracelets, the more tortured the coiffure and ultra chic the hat, the less she seemed really a woman. In general, she did not know how to please a man and so either remained single, or married a failure whom she militantly protected. The long, nacreous fingernails were bayonets barely disguised.

In a kind of senseless effort to disassociate myself, I began wearing old tweeds and country walking shoes to fashion meetings. I wasn't sure exactly what I wanted to do, but I knew it was time to leave.

For some time, the editor of *Stage* Magazine had been making tentative overtures: "We need someone just like you . . . part time. Could you suggest anyone?"

I had tried earnestly to be helpful but no one whom I suggested seemed to fit the job, which was fortunate because suddenly I realized that I wanted it myself.

I had a long talk with the editor; the magazine was subsidized by the same people who owned the then ten-year-old, fabulously successful *New Yorker*. I did not think it was necessary to mention my initial misjudgment on that score.

The specific work involved three pages which were set aside for fashion in a national theater magazine. There had to be some tie-in, no matter how tenuous, with the theater. In general I would be allowed to express my own ideas; I could use copy or photographs as I saw fit, but I would be obliged to concern myself with the more important shows on Broadway.

Obliged? I felt a wonderful sense of excitement rising within me; it was like the day before spring.

"When would you like me to start?" I asked the editor.

"Yesterday," he smiled.

It took a little while before I could effect the change. The

Search for Identity

day I moved into my new, sparsely furnished office with no view, I found a little plaque on my desk: my name followed by ASSOCIATE EDITOR. It had taken me ten years! I ran my hands lovingly over the cheap, yellow oak desk.

"I think I'm home," I breathed, "and for *good!*"

8

Still on the Train

Inexplicably, the current of my reminiscences was cut off. I was graveled for further recollections. Night had fallen and the train tore through the blackness, hooting a sharp warning to itself as it neared a bend. I swayed and when I put out a hand to steady myself I was surprised to find that the shiny oak desk was not beneath it.

"That's about the story," I said to Howard. "You know it from there on."

"I should," Howard said softly, "we were married a couple of months later. Why didn't you ever tell me how much that job meant to you? We could have managed so that you could have kept it."

I shook my head. "It seems to me I've been making changes

Still on the Train

all my life. And that job—like everything else I've ever done—was right, but in its time. And then came the time to open my hand and let it go, otherwise there wouldn't be room for anything new. And, at the moment, I feel as if I had everything!"

Howard acknowledged the tribute in an appropriate manner.

"I'll tell you what," he said a little later as he was removing a smudge of lipstick, "we won't define *how* you pick plays. Just keep right on, the way you've been doing."

The next morning our train pulled into Chicago, where it would lay over for four or five hours before continuing the westward trip. The train trip had much of the atmosphere of an Atlantic crossing: friends would come down to see one off and everyone was very careful not to get too chummy with anyone the first night—one waited to see who got off and who got on in Chicago.

We had already encountered an old friend, Brooke Marshall, who was taking along her small, wiggly dachshund, Siegfried. Brooke was somewhat apprehensive about giving him his exercise at the early morning stops; but Howard, who was immediately captivated by Siegfried, promised to spell for her.

In Chicago we lunched at the Blackstone with friends and encountered more acquaintances who were taking the same train. Back at the station, the sense of a sailing was heightened and we walked up and down the platform, getting a final bit of exercise before the long confinement. I nudged Howard: there was a man garbed in a plaid cape and purposefully striding up and down the platform as if it were an English moor.

"A road company version of Hawkshaw the Detective?" Howard hazarded.

"Only from the neck down," I said. "He has a head like a clever, forty-year-old, blonde cherub."

The whistle blew. "Come into the lounge car," someone

Occupation: Angel

urged us. "We're having a little get-together."

The lounge car was more crowded than I had ever seen it. Our friend greeted us; next to him stood the blonde cherub, who was introduced as Mr. Evelyn Waugh. No reigning monarch and no combination of the three most alluring male stars in Hollywood could have found me more worshipful. I remembered some of Waugh's books I had loved: *Vile Bodies, A Handful of Dust, Mr. Loveday's Little Outing*, and his most recent, *Brideshead Revisited*. I thought of the many hours of almost exquisite pleasure his work had given me and I wanted to offer him some perfect tribute—most of all I wanted to offer it in such a brilliant and witty manner that I would be memorable. But how does one impress a man who has been described as "the only first-rate comic genius that has appeared in English since Bernard Shaw"?

Whatever the secret formula might be, I didn't have it. My mind felt as thick as my tongue and I barely managed: "I've never forgotten Mr. Loveday or Bella Fleace. Any time I think of them they can make me laugh, or want to cry. Thank you for them."

In retrospect, I suppose that to a writer—and perhaps to anyone—a simple tribute is more appealing than sagacity. Or perhaps, the fact that I was familiar with two of his lesser known characters automatically imbued me with an appreciation bordering on brilliance. In any case, Mr. Waugh steered me straight to a chair with a comfortable arm which he occupied. We progressed from the old books to the current *Brideshead*. He appeared surprised when I said that I already had read it. I quoted a line or two as proof and explained that I had gotten hold of an English edition before it was published here.

For the time being, at least, Mr. Waugh was lost to the rest of the party. He told me that *Brideshead* was responsible for his trip to California. A moving picture company was inter-

Still on the Train

ested in buying the rights but Mr. Waugh wasn't sure that he wanted them to do a film of it. They had offered to pay all of his traveling expenses and he was on his way out to see them and talk it over.

"They're quite insistent, moving picture people, aren't they?" he observed. "And when they want you, they want you right away. Actually, this wasn't too convenient a time . . . we have a new baby . . . haven't really had a good look at it."

I felt suddenly protective and aghast at the moving picture people who were pushing him around that way. I told him that he should have held out and arranged to take a later crossing. The very idea of not having had time to get a good look at his own child!

"Oh, *that*," he observed mildly, "probably was my own fault. I wasn't all *that* rushed; the pram was pushed past my window often enough, I just never got to look in."

At that interesting moment, our host came over and said that no one else had met Mr. Waugh and he'd have to break up our cozy chat.

"Don't go away," Mr. Waugh urged me over his shoulder, but our contact was broken.

On the way back to our compartment Howard and I encountered Brooke Marshall, who wanted to know if we had met Evelyn Waugh. She was crestfallen at having missed him.

"Do you think," she asked hopefully, "you could possibly ask him for a drink before dinner? I've always wanted to meet him. I have a beautiful, big bottle of champagne, a jar of *foie gras* and a box of crackers—going away presents."

"I daresay—especially if Peggy asked him," my husband said pointedly. "They sat in the corner talking to each other almost the whole party; apparently they have a couple of English friends in common . . . anyhow, nobody else had more than a word with him."

Occupation: Angel

Howard went off to send the invitation via the porter and Brooke looked at me with surprise and interest.

"Oooooooo! Having a bit of a flirt?" she wanted to know.

"No," I admitted somewhat reluctantly, "it was just because I loved his books. And they weren't friends we were talking about, but characters from his stories."

And then, because I couldn't resist it, I told her briefly the story of *Bella Fleace Gave a Party:* Of the eccentric and very, very old maiden lady who lived in a derelict Irish castle and who hit upon the idea of selling some valuable first editions and splurging with a great ball for all the neighbors who lived within a hundred miles and whom she never saw. She spent months in the preparations, which were stupendous. New servants were brought up from the village to scrub and polish; paperhangers, painters and plumbers got to work; windows were reglazed, rugs were shifted and the furniture was moved; unused silver was polished and long-forgotten porcelain services were found. Despite her ninety-odd years, Bella was indefatigable and she had a joyous time ordering herself an elaborate ball gown, deciding on an orchestra and a caterer, choosing the menu, planning the floral arrangements and, best of all, deciding upon her list of guests and addressing the great stack of parchment envelopes and engraved cards.

But the night of the party, the candles flickered over the empty rooms and the orchestra echoed hollowly. None of her expected guests arrived. The disappointment was too much. The frail old lady collapsed and the next day she died. The English nephew and heir whom she disliked came over for the funeral. Afterward, he was sorting out her effects when he found in her desk the huge pile of stamped and addressed, but unmailed, invitations to the ball.

"I can't do justice to it," I protested, "it's the *way* he writes it, so that it's funny, and sad; and once having read it, a picture of the castle and the way it looked that night stays in your

Still on the Train

mind forever."

Howard returned with the news that Mr. Waugh would be delighted to drop in at seven for a cocktail before dinner. Brooke thanked me for briefing her and dashed off in a flurry to have the champagne iced and to change for dinner. At half past six Brooke was still in a flurry: her back hair did not suit her; the *foie gras* had to be spread on the crackers; Siegfried had to be walked at the next stop and he had to be fed *first* if the walk was to be effective.

The train was slowing up in anticipation of the stop.

"I'll feed Siegfried," I volunteered. "And Howard has already promised to walk him."

Brooke seemed relieved. "I've opened his little can of dog meat, but it needs to be squished up a bit . . . sometimes he's awfully picky about his food."

I messed up the ground meat. It smelled delicious and my charge gobbled it up in no time at all. He was still licking his lips greedily when Howard came to take him out.

"Now," I said briskly, "I'll tell the porter to bring the wine and some champagne glasses into our room and I'll spread the crackers."

I felt very efficient and all, including Brooke's back hair, was in order when Mr. Waugh arrived. Howard poured the champagne, which was dry and well chilled, and I passed the canapés. I had passed them twice before I got to taste one myself. I thought it rather coarse and lumpy but I ate it as did everyone else. I had trouble keeping my attention on the conversation. There was something familiar yet strange about the canapés. They did not have that lush, creamy richness one might expect; the spread was not just bland, there was a total absence of flavor. I glanced at Mr. Waugh but he was munching away happily. I looked over at Brooke; her expression was a mixture of horror and incredulity. Well, it isn't as bad as all that, I started to say to myself when everything suddenly came clear. Brooke had caught on first; in my smooth efficiency I had mixed the two jars—we were eating Siegfried's dinner and the repast he had gulped so happily must have been the *foie gras*. After I was sure of the mistake, the canapés seemed even less palatable, but I made quite an elaborate ritual of eating another one. Apparently in the empyrean in which Mr. Waugh's mind dwelt, there was no room for such petty considerations. But I had lost our earlier rapport. I sat there transfixed, watching him chew and swallow, following the curious little bob of his Adam's apple as if it might suddenly stop in protest at my culinary outrage. I, who had yearned to offer him homage, wound up feeding him dog's meat.

The next morning, about eleven, I tucked a play under my arm and walked back to the deserted club car. I had barely started the first scene when I was joined by Mr. Waugh, who appeared genuinely pleased to find me there, slipped into the adjoining chair and asked me to join him in a drink. Involuntarily, I glanced at my watch—it seemed to me an odd hour at which to start drinking but I was embarrassed and fearful of seeming censorious, so I muttered something inane about liking to wait for the sun to be over the yardarm. This was pure pretentiousness because ordinarily I didn't drink at noontime either.

"But *you* go right ahead," I said heartily, still involved in a compulsive desire to link myself with his tastes.

"I'll wait until *you're* ready," he said, and waved aside the steward.

We started to talk about Hollywood; it was clear that he really was searching for a clue to the customs and behavior of the moving picture people.

"I suppose you've been out here often—know all the people connected with the movies," he said a little wistfully.

I explained that our connection with the movies was a bit backhanded and had developed through our interest in the theater. I had the feeling that an actor who once had been on Broadway always had an urge to return either as an actor or perhaps as a producer. I went on to say that, as an example, both Robert Montgomery and Elliott Nugent were successful in the movies and living in California, but they were talking about co-producing on Broadway when they found the right play. Elliott was not only a good actor but a skilled director and also the author—or at least co-author with James Thurber—of a first rate play, *The Male Animal*. We had been drawn together naturally, and had become good friends. I explained that we referred to people like that as not being at all Hollywood. It was harder to find un-Hollywood people the higher up you got in the production end of the business, but with a little searching . . . However, I admitted that, although this was our third or fourth trip, we usually spent most of our time out in the desert and had had relatively small association with the movie world.

"Of course," I added, "we've seen movies in production both inside and on location; we've lunched at the studios and met many of the stars; we've had all the usual treatment including the standard, this-should-impress-you dinner at Mr. Louis B. Mayer's house on the top of a hill."

"Why were they trying to impress you?" Mr. Waugh asked.

"I don't know," I admitted, "maybe just to keep in practice. They seem to work on the theory that they will be better able to negotiate a deal with you after they have left you stupefied with their lavishness. I must confess we were more bewildered than impressed. All of the ladies wore rather elaborate evening dresses but none of the men wore dinner jackets, in fact, some of them wore such peculiar and flashy sports coats that it looked as if two entirely different parties suddenly had merged. The house was surprising too—a combination of massive formal Spanish furniture, great, wrought-iron gates and California casual informality. The butler was overintimate with the

Occupation: Angel

guests. Some of the guests were shockingly late—no one seemed to notice, or apologize and we spent an interminable length of time over cocktails before dinner...."

Mr. Waugh seemed at once fascinated and repelled. "I've already been invited to a dinner—probably the counterpart," he said, "but how does one cope with it? What shall I do in defense?"

"I don't know, but I've always thought it would be interesting to act blatantly unimpressed—pretend to expect more than they possibly could provide. It might drive them a little bit mad as they keep renewing their efforts. . . ."

The worried frown disappeared and Mr. Waugh's face looked more cherubic than usual. He beckoned to the steward and then once again urged me to join him in a drink. By this time it was about twenty minutes of twelve; it seemed stuffy to insist upon waiting the extra few minutes for my false drinking hour to begin—and rude to allow him to defer to me a second time.

"Perhaps a little brandy," I suggested. "A very small one."

"A pony of brandy for madam," Mr. Waugh said crisply, "and a double orange juice for me!"

I sat there stubbornly sipping my brandy and feeling alternately decadent and self-conscious while Mr. Waugh delicately drank his orange juice. It was the second time in two days that I had been the means of creating a little private joke for him. I was not exactly the butt of the joke nor was I a partner in it; rather I was an instrument through which he created a situation which he found diverting and slightly ridiculous. If I happened to pick up his wave length of humor, fine, but he remained innocent and slightly remote.

He urged me to tell him more about Hollywood—what was there to see, besides the studios?

The queer, the ridiculous things, I assured him, he could barely miss. They would confront him, surround him; the

Still on the Train

manmade ugliness and the natural, lush beauty. I told him about the roadside stands fashioned in the shapes of the produce they sold: the two-story ice cream cones and the long, revolting frankfurter houses for dispensing mustard-smeared little prototypes.

"And if there's a car at your disposal," I continued, "don't miss the pets' cemetery. It's huge, beautifully landscaped and has elaborate headstones, mausoleums and even marble statues. The graves are smaller than the kind we are accustomed to because they contain dogs, cats, and even canaries. I hear they have an elaborate funeral home where services are conducted, but I can't swear to that—I've only had a quick glimpse of the place."

Suddenly, I was graveled for new ideas and I wondered if it was because Hollywood offered such a paucity of choices or because I knew instinctively that the usual sights would not interest Mr. Waugh. He said quite solemnly that he would remember everything I had told him about the place and the people.

A few years later when I read *The Loved One* which Waugh himself described as "a little nightmare produced by the unaccustomed high living of a brief visit to Hollywood," I knew that he had paid a visit—and perhaps many visits—to the animals' Valhalla. But within two weeks I had another testimony that he had listened to my description of Hollywood. And this time I had the strangely privileged sensation that he had written me a thank-you note, a message on the wind.

We were in Palm Springs when it was innocently delivered.

Each time we went to California we met one of the producers at Metro-Goldwyn-Mayer who vacationed with his wife and children at Palm Springs. He was an Englishman, or more precisely, an Australian, who had made California his home but whose accent instead of diminishing seemed to increase with the years. This proclamation of British background plus

Occupation: Angel

his flair as a storyteller gave the producer an enviable standing in the company.

He sought us out at the Club almost immediately upon his arrival and, after the first amenities were exchanged, he announced that he bore greetings from a *very* close friend of mine. I had trouble guessing.

"Evelyn Waugh, of course," he exclaimed. "He tells me you came west together."

I would not have presumed that the three-day train trip would have established us as old friends; nevertheless, I was pleased that Mr. Waugh had referred to me in such a manner.

"Ah, yes, Evelyn Waugh." I repeated his name because for the moment I could not think what else to say; then I rallied and added inanely: "Charming fellow."

This was the Australian's turn to be silent for a moment.

"Y-e-s," he admitted almost grudgingly, "charming but, well, I wouldn't want to say a word against him, your being such old friends and all, but to tell you the truth, we found him a bit, er, a bit *difficult*."

From my own brief encounters, I decided that well could be the understatement of the season. But aloud, I merely urged: "Difficult? Tell me about it."

"Well, to begin with, we wanted to do something nice, of course, and we had planned a party for him . . . a *big* one at Mr. Louis B. Mayer's own house." There was deep respect in the producer's voice. "You know the place . . . you've been there to dinner . . . that long, private road, the big gates, and the house which is, to put it mildly, rather, er . . ."

"Impressive?" I suggested.

"Exactly," he agreed. "It's a pretty impressive setup and we had some of the most important people there to meet him."

"Surely," I said, "he must have been appreciative of all that trouble. He wasn't *rude*, was he?"

"No, not rude," the producer hastened to assure me. "At

least he didn't *mean* to be."

"But what did he do?" I probed, and my interest was not feigned.

"It wasn't what he *did*, it was what he *said* and right in front of *everyone!* He looked around him and said:

" 'How wise you Americans are to eschew all ostentation and lead such simple, wholesome lives! This really is delightful. Who'd even *want* to live in the *main house* when he could have this charming *gate house*, instead!' "

It seemed to me that the moisture on the producer's forehead was the result of the recollection rather than the desert heat.

"You could have heard a pin drop! None of us could think of a thing to say. He didn't seem to even notice it, he just stood there with a pleasant smile on his round, simple face. I can tell you, it was all rather unnerving. Now we don't quite know how to deal with him, even on a business level. . . ."

A sudden recollection of our supposed intimacy smote the producer:

"Of course, everything I've told you is in the *strictest* privacy. You won't ever mention any of it to *him?*"

"I promise. Probably I won't even see him before he goes back," I said, as if it were a rare and almost insupportable separation. "But somehow I don't think he'd really mind my knowing. He'd know I'd understand."

9

Angels with Clipped Wings

Of all the aspects of the theater, angeling might seem the most remote and removed from personal involvement. It didn't turn out that way for us. We were investing in ten or twelve plays a year and somehow my search for these plays and my concern with them was occupying more and more of our time.

I really meant to follow Howard's advice to continue selecting plays in the same way I had been doing; it just wasn't possible. Instinct is a handy adjunct but, in the long run, it is not enough. It was inevitable that I should begin to learn the rudiments of a trade that was already my business . . . big business.

In studying and analyzing our successes as well as our failures, I suppose it was equally inevitable that my approach

Angels with Clipped Wings

to selecting a play lost its initial simplicity and consequently much of its conviction. I no longer could tell when my divining rod was working automatically or when it was being nudged by empirical knowledge.

Unfortunately, we already had gained a reputation for great sagacity in the theater so that this was an awkward time for me to confess how little I actually knew. I was still in the early process of learning. However, any admission of doubt on my part was interpreted either as excessive modesty (in a world where modesty is highly suspect) or an unwillingness to cooperate and share our success formula with others.

So usually I'd suggest to a would-be backer that he start with $30,000 or $40,000 that he could afford to lose; invest it in five or six good plays to spread the risk; and then invest the profits in some more good plays.

This couldn't be a simpler recipe . . . or more difficult to follow. And it ignored the outside factors. But then it took me quite a long while to catch on to those outside elements myself, and to estimate when they should influence my decision on an even level with the play itself.

Learning to read and evaluate a play is a little like learning to write one.

As you read a play, you must see the action projected in your mind's eye. And as you see it, you must endeavor to react both as an individual and as an audience.

I think the reason why the average actor's estimate of a play is often the least sound and valid is because his approach is purely personal. He sees one character—the part he would like to play—brilliantly illuminated against a fuzzy background.

I once asked an actor how he liked a play he had just finished reading.

"I love it!" he answered without reservation. "It's beautiful . . . and I'm on in every scene!"

To a certain extent, every reaction is bound to be personal

but unless the reader also has a feeling for the mass audience, a sense of the least common denominator, his opinion is useless.

No matter how daring, how swift and exciting the author's flight of fancy, there is the ballast to consider: can the play carry along the audience? An audience which averages 960 different people a performance, eight performances a week? An audience which in only one season of forty weeks adds up to 307,000 individuals, united under the generic term "the public"? This is not a dead weight to carry. Anything but! And the person who shrugs off a failure on the grounds that it was too "good" for the public is a fool. When audiences consistently and enthusiastically will support the plays of Aristophanes and Sophocles, Shakespeare and Shaw, Chekhov, Ibsen and Strindberg, Eliot and O'Neill, no apology need be offered for them. If they won't accept a play, the chances are it isn't good enough for them.

One of the many hazards of reading and judging a play is that the impact of a statement upon a person reading it by himself may be very different from its effect upon a crowd hearing it in the theater. At home one may read and consider a theory in a leisurely manner. But when one is trapped in a seat for hours at a time and surrounded by a thousand other people with their inevitable distractions, the theory must make its point immediately, convincingly and in such a way that each person in the audience can identify up to his own capabilities.

If it is true that an audience is a hazard, it is equally true that it is a natural resource capable of turning a quiet chuckle into a gale of infectious laughter.

A reader not only must see the play's story line projected in his mind's eye, he must also see—and most important *hear*—the actors. Are their entrances and exits smooth? Plausible?

It is possible that the most difficult skill is learning to hear

Angels with Clipped Wings

the dialogue with one's eyes. Oddly enough, there is a world of difference between dialogue which is effective when read and that which actors can speak. An accurate and precise mind's ear is needed to recognize the difference.

One or more of the characters must catch the imagination of the reader. One may loathe a character and yearn for him to get his just deserts; or one may feel an immediate empathy and hope that a solution of his problem can be found, but one way or the other the reader (and eventually the audience) must feel involved and personally committed. Otherwise, the wittiest dialogue, the most hysterically funny situations will count for nothing. A good yardstick to apply is: Who do I care about?

To become a reasonably good judge of a manuscript takes an odd combination of instinct and experience, intuition and love. But even then, without the addition of magic, the rest doesn't count for much.

I believe that magic can enter only when the groundwork is prepared . . . and the play itself is the groundwork.

Unfortunately, it is *only* the groundwork. And even if for one blinding moment the absolute could be seen and one might say with assurance, "This, in its present and exact form is a perfect play," it wouldn't prove anything because a play in manuscript form is little more than an embryo.

Plays are meant to be acted and before the manuscript's suggestion of life is brought into a quickened production, a long and dangerous road must be traversed. And each step of the way may lead, independently of the script, toward success or failure.

Other art forms generally are born of one person. A play is conceived by one person, born of many. And thus it is given the advantages of many talents, and all of the problems.

The part of the leading lady might have been written with a specific, outstanding actress in mind but perhaps for one

Occupation: Angel

reason or another she cannot accept the part. A second choice will have to be made and everyone agrees that it must be made among name players.

The first candidate for substitution, although a good actress, has no name; the second has a following and surely could extract the humor in the play, but she has not the dramatic depth and range that the part also requires; the third has both the name and the dual talents needed but, alas, she is a good fifteen years too old.

Here creeps in the first compromise and possibly the first substitute for perfection. From then on each bit of casting, each piece of scenic design, each line of directing may enhance or destroy the original conception.

Sometimes the personal chemistry of the people involved may work against the play's success. A good, adaptable director may find himself inexplicably incapable of communicating with the leading player. Or the two romantic leads simply cannot strike a spark together. And no declaration of passionate and deathless love will convince the audience that this man and woman are anything but bored with each other.

A brilliant example of this, but in reverse, was the casting of *South Pacific*. Both Mary Martin and Ezio Pinza had outstandingly successful marriages, but when they faced each other on the stage they were able to project a sense of such naked emotion that even blasé Broadway audiences were half convinced they were seeing something not intended for them. At eight performances a week, a percentage of the audience left feeling that the two leads were involved privately in no mere flirtation of the mind but in a deep and passionate love.

For one reason or another—sheer infatuation, I daresay—I saw the play eight times and at each performance, when Mary Martin and Ezio Pinza looked at each other, I thought I saw the whole stage light up. That is the rare plus that every author dreams of but rarely sees achieved.

Angels with Clipped Wings

Some plays have failed through sheer chronological error. A good play with a convinced pacifist for a hero would have been no less a perfect play if it had opened the week after Pearl Harbor . . . but it would have stood little chance of success.

After I came to understand some of this, I conceded to myself that the actual manuscript in hand was only part of the picture. Quite understandably, the young playwright is apt to ignore this and bitterly insist that producers and angels lock themselves up in ivory towers dreaming of great names and ignoring the newcomers. This is, of course, untrue. Basically, one looks for a *good* play but there is no doubt that it is more attractive for it to be by a skilled, prolific and experienced professional writer rather than an untried one. This preference is not based upon any arbitrary or snobbish urge for a name. Nor is it because the public will pay homage to a writer the way it does to a famous star. With the exception of half a dozen big names, the public is coolly indifferent to the author; his name is barely attached to the exploitation of the show; and, as the audience leaves the theater, it is doubtful if one person out of fifty could tell the name of the playwright.

The producer recognizes the fact that even with a good play, the difference between eventual success or failure may depend upon the author's ability to sharpen a scene, cut or even rewrite between the time the show starts its out-of-town tryout and opening night on Broadway. Although the advantages of working with a seasoned writer are obvious, producers are not so shortsighted as to be unaware of the need for developing fresh talent. The question is how? Certainly not at the current rate of $100,000 a lesson. Yet a playwright learns only by seeing his mistakes in action. Maxwell Anderson used to say that a successful first play was always largely accidental.

The problem is a two-edged sword. And if the young writer who never had a chance to try out his plays back in the $15,000

Occupation: Angel

to $20,000 bargain days is out of luck, so is the serious producer who knows that he is facing an ever-narrowing circle of talent. He realizes that any season offering a disproportionately large number of revivals not only presages a sterile future in the theater but yields limited returns, and consequently a smaller capital with which to try an experimental work.

The path of a young, creative worker in any field is difficult —the playwright's is no worse than anyone's else. And the well-known playwright is in a much better position than proportionately successful poets, artists, and writers in general, although he justifiably may feel that of all forms of art, his is the least secure, the most likely to be tampered with.

A piece of music stands or falls on the composer's conception; and a nude chiseled out of marble won't take draping. A finished painting is a completed entity; it may be criticized, praised, bought or rejected but no one is going to suggest that he will buy it if the artist makes the roses anemones, or uses a deeper, more vibrant blue on the bowl.

But the author who finally achieves a production rarely finds that the play produced on Broadway is identical with his original manuscript. Indeed, based upon a later, careful comparison, one might come to the conclusion that the original script represented little more than the basis for negotiation between the producer and the playwright! This certainly is not sheer captiousness on the part of the producer. Neither he nor anyone else can be absolutely sure of dialogue before it is spoken or action before it is played.

Sometimes the changes necessary to a play's success may lie within its own construction; sometimes the alterations are based upon a star's special talents which must be exploited or his limitations which must be covered.

The producer-author relationship is both difficult and tenuous. Surely it can be no accident that in a comparable relationship, the average novelist goes back again and again to his

Angels with Clipped Wings

same publisher while a playwright is likely to make constant changes, always searching for the very things no producer can promise: success or perfection.

As backers, we found ourselves practically in the position of umpires in a game where there are no sure rules. But at least we learned to see both sides of the question.

The initial relationship between an author and the producer who has just agreed to put on his play may bid fair to rival that of Damon and Pythias. The writer feels understood. The producer feels appreciated. They love and trust each other but legal steps must be taken to put the production in work. This is the moment when the age-old picture of the helpless artist in the hands of the ruthless money grabber begins to dim.

As a member of the Dramatists Guild, the playwright will expect to use their standard contract. However, before this contract may be signed, the producer first must sign a quite separate contract with the Dramatists Guild; and this, surprisingly enough, stipulates that any *other* writer the producer signs up also must be a member of the Guild or join it within thirty days.

If, in defiance of this rule, a producer should put on a play by a playwright who was not a Guild member and did not promptly join, the producer would be blacklisted by *all* the Guild members! It also is likely that brother unions such as Actors' Equity and other theatrical guilds would join in the boycott. This would seem a direct violation of the Taft-Hartley Act, which specifically denies the right to such a secondary boycott. But no producer dares to risk an open issue on the subject, and so he signs this contract.

Next he and the author jointly sign a pledge which obliges the producer to stage the play within one year or "his right to produce the play and to the services of the author shall then automatically and without notice terminate."

The producer also promises to provide "a mutually agree-

Occupation: Angel

able, first-class director . . . and cast . . . and to present the production in a first-class theater . . . in a first-class manner . . . and in a regular evening bill."

At this point, it may seem that practically everything is guaranteed to be first-class except the play itself.

Physically, these contracts are lengthier than the most complicated will, and run to thirty-odd pages each.

In effect, the second contract promises the author final and complete control over his script, veto power over the producer's choice of a director and over the producer's and the director's joint selection of casting.

Usually differences of opinion can be met and ironed out but, should the author prove adamant, the producer must agree or abandon the whole project.

Upon the signing of the contract, the author gets an advance of at least $500 and the guarantee that, based on the gross, he will receive a percentage roughly the equivalent of 10 per cent. Since a straight comedy that proves a hit may take in about $43,000 a week, the author stands to make on the gross of receipts $4,300 a week—a tidy sum by any standards! Also, this percentage is the minimum guarantee; a successful playwright may demand even better terms. And, unlike the producer or the backers, he gets his money right from the beginning, before the initial cost has been repaid. However, an obvious hit yielding over $40,000 a week should be in the black, yielding a return to producers and backers alike in a relatively short time.

If the play fails, the producer has risked the loss of his backers' good will and confidence; his reputation has been somewhat diminished; and he has worked half a year for practically nothing. Certainly a play that is unsuccessful financially is not to be desired by the author; if it closes out-of-town or is an immediate failure on Broadway, he may see nothing but his $500 advance for one or two years' work. On the other hand, it is perfectly possible for a play to limp along for months,

Angels with Clipped Wings

fail to pay back its original cost, yield little for the producer but his office expenses, nothing at all for the backers and still show a profit for the author.

A few years ago, Alec Coppel wrote a play called *The Gazebo*. It ran for six and a half months at little more than break-even business. When it closed, despite a $90,000 moving picture sale, the backers still had a loss of over $50,000. But the playwright had *made* $54,000 from the movie sale and $60,000 royalty on the Broadway run, a total of $114,000 on a play that was a financial failure!

It is no wonder that the author and producer see the financial aspect of a production from diametrically opposite sides. The author's chief concern is the gross, the producer's the eventual profit.

One of the many financial problems a producer must face is whether or not to sign up for theater benefits. If the show has a famous author, composer or star it will be sought after. There was a time when successful producers would flatly refuse theater parties; now there are few who dare to decline the assurance they give. The theater benefit has grown from just a way of raising money for a favorite charity into a professional business grossing $8,000,000 or $9,000,000 a season. Each benefit assures a complete sellout at box office prices (minus the 5 or 10 per cent given to the theater party broker). If the producer has any qualms about his show—and who hasn't—it is a comfortable feeling to know that two or three benefits a week for the first month may keep the play alive long enough to build by word-of-mouth advertising, and possibly help to surmount mixed critical reviews. However, two or three benefits a week will not cover the full running expenses of a show and, if it looks like a dud, the producer may have to cancel the benefits, return the deposits and close the play. If it is a smash hit, the producer may regret his bondage to the benefits and the needless loss of the percentage paid to the

Occupation: Angel

play brokers.

Theater ticket brokers and scalpers resent benefits as an infringement upon their natural rights, with the extra money going on a tax-deductible basis to a charity instead of to them; box offices protest that benefits complicate the mail order business and antagonize the potential theatergoer; actors openly deplore them; and directors complain that it may take several performances before the show is back in its natural groove. People who attend benefits come from every level of society, and, although each person may have marked individual characteristics, he almost automatically becomes part of a difficult mass personality characterized by late arrival, talkativeness, an ungenerous approach to the performance, and a social manner better suited to a party than an evening in the theater. Unlike the any-night audience, which usually arrives in twos, eager to see that particular show, the benefit audience bursts in in groups, fighting a losing battle with curtain time. Fresh from dinner parties, they chat in tight little knots in the aisles; and keep an eye out for even later arrivals whom they greet with waves or noisy whispers. Some of them seem openly rebellious at the prices they feel they were pressured into paying and maintain a tight-lipped, see-if-you-can-give-me-forty-dollars-per-ticket's-worth-of-fun attitude! Sometimes the tickets have been bought to please the wife of a business associate and the purchaser is barely aware of the show's theme. There was the man who was bitterly and noisily disappointed to discover that *Man and Superman* was not a dramatization of his favorite comic strip; and I recall one matron who stalked out of *A Streetcar Named Desire* observing indignantly that the Speedwell Society's morals were slipping.

Whatever the artistic liabilities may be, the fact remains that a guaranteed sellout of $5,000 or $6,000 for a straight show and $9,000 or $10,000 for a musical add up to a good deal of financial insurance. One much-touted musical opened with

Angels with Clipped Wings

over a hundred benefits booked—nearly $1,000,000 with which to soothe ruffled feelings.

There exists a strange chain of interdependence which not only links the producer and playwright but also the producer and the angel. And they alternately are bound together with the devotion born of a financial success, or torn asunder by the blame-slinging that so often comes of failure.

One has a right to expect from a producer a rare combination of talents—including honesty. These qualities are not easily determined upon slight acquaintance; only experience, and sometimes repeated experiences, will give the answer. Producers are like Tuesdays . . . sometimes good, sometimes bad and sometimes both.

However, the playwright has less need than the backer to be concerned about honesty; he is protected by one of the strongest unions in the country. He is not even bound by his contract until he has received a down payment in advance against future royalties; and as these royalties come out of the gross receipts at the box office, no matter how extravagantly or prudently and honestly the producer may work out the running costs, the playwright gets his first.

There is no similar, protective union for angels, and so we are more vulnerable. Technically, we are protected by law, but then technically we are protected against common robbery, too.

Once upon a time there was a producer who was not without a certain standing, although he had not had a hit for quite some time. He offered us a play; I liked it, and we took a share and sent a check. Perhaps the play was not as good as I had thought or perhaps the other prospective backers lacked vision; in any case, he couldn't raise the rest of the money. Technically, he couldn't touch our money until all of the funds had been raised, but in actual fact, he had been living on it; not whooping it up, just paying the rent and the butcher, working

Occupation: Angel

and hoping to raise more money and eventually pay it back. We could have had him sent to jail, but that would have been a profitless arrangement for all concerned. I couldn't even work up much moral indignation—neither could Howard.

Strange, then, that we both should have felt so violently outraged when we were cheated in a successful play!

In the second case, it was an even better-known producer, creative, imaginative and talented. He offered us a play which was charming, gay, and sprightly but *light!* In the right hands it could be made enchanting, but this was clearly a case where stars were needed. They could be had—for a price. This meant a smaller return on our money, but we went ahead. And even the stars invested money of their own in the show. It was a hit and we all congratulated each other and relaxed in the warmth and bright ambience of success.

Rave reviews of this play were scarcely cool before the producer had a new and even more ambitious plan, this time for a musical. I read the story outline and, although generally I insist that this is an unfair way to judge a musical, I knew that I had gone far enough. Neither songs nor choreography can create a story, they can merely supplement it or extend it and in this case I thought to extend the story line would be to compound the original disaster. I declined; so did several of the other backers.

We weren't exactly trying to count our chickens before they hatched, but we did feel that it was time for us to begin counting a few shekels from the hit. It was difficult to get the usual financial statement. But that, we were told, was because the producer was so busy with his new show.

Eventually the new show opened and closed almost in one smooth motion. Then we found out where our money had gone (and "gone" is used justifiably)—into the new and promptly late $450,000 musical!

"Apparently, some producers just won't take 'no' for an

Angels with Clipped Wings

answer!" I observed to Howard.

"I don't care what it costs me, I'll sue—and I'll keep on suing, if necessary, until I get the money back!" Howard swore.

In my own way, I was quite as irritated as he was but I could not help but wonder why one dishonesty filled us with sympathy, another with rancor. Was it because the latter seemed such a gesture of arrogant defiance? And I wondered what would have happened if the musical had proven a fabulous hit: Would the producer quietly have *returned* the money, using his intense preoccupation with the new play as an excuse for a thoughtless delay? Or would he possibly have announced to us that we were the co-owners of part of a new gold mine?

I doubt if we'll ever know. We sued and won our case. We collected $10,000 out of the $19,000 he owed us and got a note for the rest. Not only did the producer show no sign of embarrassment but later, when he was planning a new production, he called up and genially offered us a piece of it. He even made what I am sure he considered a gracious gesture: an $11,000 participation in exchange for our $9,000 note.

We politely declined. With equal politeness, he accepted our decision but as he rang off I could sense his murmuring: "What a couple of spoilsports!"

Another time, many years ago, a play that might have been classified as artistic was submitted to us. I was fascinated with the writing and delighted with the tentative casting that had been lined up.

"How much of the play is still open for financing?" Howard asked the company manager.

We were aware of the fact that backers were proving hesitant and that the gossip on Broadway was that the play would be another artistic failure. And so we were not too surprised when the company manager admitted that they had been able to raise only $30,000 out of the $60,000 for which the play was capitalized. Howard said we'd cover the whole deficit. The

Occupation: Angel

company manager thanked him and hung up.

A few weeks later when the contracts came in, Howard was confused to see that we were down for $5,000 instead of $30,000. Meanwhile, enthusiasm for the show had mounted and rumors of a hit-to-be were spreading around Broadway.

Howard telephoned.

"There's a mistake in my contract. It calls for $5,000 and you will recall that I said I'd take the whole $30,000 you were short."

"That's right," the manager agreed, "but do you know something? Your names worked magic; every time I said you were taking a large share, the money came rolling in—no trouble at all. But it did mean that I had to cut you down. I know you won't mind."

"But I mind very much indeed!" Howard said. "I made a *firm* commitment!"

"Oh come now, Mr. Cullman, I had no idea you'd take it like this. Really!" The manager's attitude was that of a patient teacher who almost has come to the end of his good nature with a fractious, demanding child.

"I can tell you one thing," Howard said, somewhat irrelevantly. "If you were on Wall Street, you could fail to recognize a commitment just once and you'd be thrown off the floor of the Stock Exchange before you knew where you were at."

After they had hung up with equal coolness, Howard rubbed his forehead in confusion:

"I know how to deal with a crook—if *he* knows he's a crook—but I'll be damned if this fellow doesn't think he's behaved in a perfectly normal, honest fashion!"

We learned from the incident that our names carried weight. Nineteen out of twenty backers were winding up each season in the red and some of them were finding it easier to try to follow the twentieth team which obviously was making money. The knowledge was flattering but it bothered me, although I

Angels with Clipped Wings

wasn't sure why. Then, a few months later, a young producer explained it all too clearly to me.

He was completely forthright: he had submitted to ten people a play that we had turned down. These ten also turned it down, but five of them first had asked: "Are the Cullmans in it?"

"I suppose," he said reflectively, "that I could have said you were—or that you hadn't yet given me an answer but they might have known you and double checked on me.

"It is perfectly possible that none of them liked the script but, frankly, I think some of them are afraid to make a decision on their own. They'd rather follow you."

This was the last thing in the world we wanted. I tried to tell the producer how sorry I was—and how helpless I felt to correct the situation. But he had it all figured out:

"Take a piece of the show, I don't care how small, say 4 per cent. . . ."

I started to protest but he motioned me to silence while he continued.

". . . that 4 per cent of the capitalization should yield you 2 per cent of the profits on the usual fifty-fifty split between backers and producer. Right? But I'd be willing to give up that 2 per cent of *my* share. That way you'd get double the return, a full 4 per cent of the profits!"

(Later I learned that this is known as "one-for-one" and is sometimes demanded by actors and others who wish to participate in a play, but who feel that their money is worth more than other people's.)

I declined the offer.

"You strike a hard bargain," he said in a tone of real respect. "All right, I'll make you a *present* of a split share—$500 worth! And I hope it'll bring us both luck."

It was hard to explain that I couldn't accept this offer either.

"What've you got to lose?" he wanted to know.

Occupation: Angel

If he didn't know, it was too late for me to tell him, and he had a kind of innocence of integrity which was just as touching as any other kind of innocence. It destroyed utterly any sense of outrage I ordinarily might have felt.

"Let's just put it down to pride in my record," I said. "I don't want to risk lowering the score. And I'm truly sorry. I know it's too late to do you any good on this, but from now on I'm not going to talk about which plays I like or don't like, and why."

A fortnight later I read in a theatrical gossip column that the producer had dropped his option on the play and I felt a fresh wave of guilt. Within a few months, it was produced by someone else and I read the death-knell notices the next morning with a sense of relief that was shockingly close to pleasure.

I have tried to keep my word, but with somewhat doubtful success. It is hard to curb my enthusiasm and my natural inclination to talk about something that excites me. Silence in regard to the plays we don't like is likely to be all too accurately judged. To say "I don't think I should discuss it" seems ridiculously pompous. Now, in the face of a direct question about a play I don't like, I hedge and hope that "It's home on my table now" will be interpreted as "I haven't read it yet."

The trouble with reminiscing about producers is that the odd, offbeat situations stand out and the reliable, competent and dedicated producers tend to merge into a comfortable blur. Perhaps some arcane darkness in our natures makes us remember the storm and forget the calm.

10

A Star in the Kitchen

For me, the best way to approach a play is alone. I am not a sociable reader who either feels compelled to share choice bits, or enjoys nibbling the literary goodies someone else may be sampling. Generally I become so immersed in what I am reading that I don't even notice interruptions. Or, at best, a recording on the seismograph of my subconscious is indicated by: "Mmmmmmmmmmmmmmm."

This is a switch from "Ummm-hummmmmmmmm" which has a more affirmative connotation and was formerly interpreted freely by the children, who constantly swore that I had given them permission to do a number of highly improbable things. I once tried to protect myself by printing on the back of a much beat-up DO NOT DISTURB sign the pronouncement:

Occupation: Angel

THE ANSWER TO ANY QUESTION ASKED WHILE I AM READING A PLAY IS "NO." It took a while, perhaps two plays later, before they learned to outsmart me with: "I don't have to take a bath tonight, do I?"

In any case, I am willing to go to quite a bit of trouble to see that I get through the straight plays by myself. Authors often volunteer to read the play to me or to explain its true meaning in advance. As tactfully as possible, I explain that, in most cases, if I am not bright enough to catch on, a great many members of the audience may have the same handicap.

I suppose I read an average of about a hundred plays a year, searching for a precious dozen.

Top theatrical producers seem to have a disproportionately large number of scripts that require immediate reading in August and early September. But unsolicited plays arrive by mail at all seasons of the year, forming a high, slithery pile on our library table.

At first the plays came to us directly from the producers—sometimes, as in our first investment, at the suggestion of the playwright. But as the stories of our prowess as backers grew, we began getting plays to read from all sources.

Each year thousands of new plays are written; their ultimate goal, a Broadway production. They come from every hamlet and town, from people in every walk of life, and they flow into New York like countless tributaries feeding a great river.

It is doubtful if anyone could estimate the number of plays that never quite reach completion; or the ones that are finished but lie in a bottom drawer waiting for a final rewrite. But, on an average, six thousand plays a year not only are completed but are taken seriously enough to be registered in the copyright bureau in Washington. All we have to do is find the right twelve plays in six thousand. Of these six thousand, it is unlikely that more than fifty (or 1 per cent of the whole) actually will open—however briefly—on Broadway.

A Star in the Kitchen

From these figures, one might be inclined to deduce either that as a nation we have an enormous quota of professional playwrights or that a few playwrights turn out an astonishing number of plays each year. Neither conclusion is true.

Perhaps Elmer Rice came close to describing the situation accurately when he said that in our land of plenty everyone has two professions: his own and playwriting.

Apparently, to the layman, writing a play is dependent chiefly upon having several weeks' free time. The skill, the specialized knowledge, the sense of perfect timing and the years of experience needed are comparable to the requirements for most other highly demanding professions. No one ever plans to take a fling at dentistry during his summer vacation, and come home with a dandy, removable bridge; but playwriting remains the great amateur challenge. It is widely recommended to untrained writers by their families, apparently on the theory that flowery descriptions may be hard, but what could be simpler than dialogue? Let a neighbor elope with his cook and not less than three people will observe sagely: "There's a play in that!"

As a protection against being confronted with just such a play, some method for screening must be devised. Logically, the first person to read any playwright's offering should be an agent so that he can make the first rough sifting of chaff from wheat.

The name of a famous writer is an open sesame to any producer's office, but the unknown needs the stamp of approval from a reputable agent. And unless there is some other, personal contact, most producers automatically reject any material that does not come from such professional sources.

No doubt a sensible backer—but then, that is almost a contradiction in terms—should have intelligence enough to confine himself to reading what the producers send him, especially since some producers option and offer for reading three times

Occupation: Angel

as many plays as they actually produce.

But I find myself unable to follow the sensible course; I have the urge to step in deeper, to prospect personally. This vulnerability of mine must be widely sensed because we receive a staggering number of plays that have managed to elude the orderly process of first being screened by an agent or a prospective producer. The playwright may have a cousin, a friend, or any remote contact who will act as intermediary and present you with a script.

Some plays arrive by mail without any introduction and lie like unbidden, ghostly guests, staring reproachfully with eyeless gaze and mutely screaming for attention. The sensible course would be to slip them into their stamped, self-addressed shrouds and get them out of the house. But the unopened script exerts a provocative lure. Under that clean red cover—fourth from the bottom—may be the work of a genius or a half-wit.

So far, by this method, I've never found the genius. But I've been led down some interesting paths.

One such unsolicited play stands out vividly in my mind. A reasonably promising plot and excellent dialogue had led me well into the second act when I began to be disturbed by several characters who not only bore no direct relation to the story line, but whom I had not even recalled in the original cast of characters. I conscientiously checked back but neither "Arthur" nor "Sybil," who were beginning to figure quite prominently in this scene, were in the listing. On the other hand, there were four or five characters who had not made an appearance.

Sybil was a brilliant and fascinating girl—her conversation pure T. S. Eliot, in prose. But after utterly captivating two of the male characters—and me—she thoughtlessly strolled out of the drawing room onto the terrace and into oblivion. Arthur was equally attractive in his own peculiar way. Unheralded

A Star in the Kitchen

and unexplained, he strode into the distinguished setting with a volume of Huxley tucked under his arm, completely disregarded the other characters and devoted himself to a discussion of the Mondrian hanging over the fireplace, and to modern art in general.

My husband, who had been reading the evening paper, looked up to say, "How is it?"

"Brilliant in parts . . . but strange." And then, warming up, I added: "It's like an essay of Kafka done in play form." I figured that this assessment contained enough intellectual double-talk to compound the confusion and give me time to go back and try to figure out the play anew. But Howard, an otherwise highly masculine man, possesses a large dose of what is usually referred to as feminine intuition.

"You mean it stinks?"

I denied this vehemently. "I think the average theatergoer's mind may not be sufficiently complicated to comprehend it." I failed to add that my own mind was in a similar condition of simplicity. But I was determined to plow through the rest of the play; I would not go down in theatrical history as the woman who had muffed a current Shaw or O'Neill.

Howard stuck to the point of the discussion: "Well, if it's got something . . . any chance of a rewrite? Who's the author, where's he from?"

I picked up the return envelope; the name was unfamiliar but the address was not—it was that of a famous sanitarium for the mentally weary.

*   *   *

Although Howard was supposed to stick strictly to the business aspects of our partnership, he was unable to resist the temptation to "drum up a little trade." If he saw a notice in the paper that a producer whom he knew had optioned a play, he'd call up and say we'd like to read it.

Occupation: Angel

One day, however, Howard overstepped the bounds. He was home with a cold—a rare occurrence. He had had his secretary on the phone for an hour; all the morning mail had been read to him and he'd made a dozen decisions; he had read the morning paper with more care than usual and, despite my obvious lack of comprehension, had explained in detail all of the errors the government was making in its fiscal policy. Finally, perhaps to reach me on a common level, he said: "What about that new play Lindsay and Crouse are supposed to be working on? I wonder if it's anywhere near completion because, if it *is*, it would be nice to see it before we go away."

As he spoke, I could see one arm reaching for the telephone.

He brushed away my protest with: "Nonsense! It's after *ten!*"

Buck Crouse answered the telephone himself and it was clear that he had been plunged abruptly and unwillingly into a new day.

"See here, Cullman," he said with some asperity, "the only reason I'm in the theater business is so I can sleep late; and if you're going to ruin that . . ." The threat was implicit.

Howard looked hurt and bewildered. He started to mumble an apology before ringing off when Buck did a complete turnabout: "To tell you the truth, Howard, we're not getting ahead with this new play very fast; there won't be anything to show for some time. But I do happen to have something here rather special . . . a new play we've been offered for production and if you'll promise to read it right away, I'll send it over. And Howard," he said softly, "I'd like *you* to read this one—not Peggy."

When the play arrived by messenger, it was late in the afternoon and I had just finished washing my hair. Howard snatched at the play as if it contained the long-awaited news from Ghent. There was a cryptic line scribbled on a bit of paper. No mention of the probable production cost or any

A Star in the Kitchen

details, just: "This much action might call for the use of a turntable on stage. Keep it in mind as you read. —BUCK"

The play was inordinately thick; it had a clean blue cover, stiff with newness. Howard weighed it in his hand thoughtfully. "This is quite a size! And a turntable . . . that sounds expensive. Do you want to take a look?" He held out the play.

I shook my head and retired under a large bath towel. "It's *your* baby," I said. Howard tried another attack:

"Why don't you have your hair washed at the hairdresser like everyone else?"

"Because this morning when you pointed out that a cold like yours could lead to double pneumonia, I broke my date at the hairdresser to stay home with you!" Privately I was a little piqued at Buck Crouse; and eight hours of a restless man around the house had made the day seem like a long weekend.

Howard made elaborate arrangements, fluffing up a pillow and adjusting the lamp before settling down to his unaccustomed task of reading a play. I ducked my head and my hair was a wet screen through which I observed his reactions. We were reversing our usual roles.

After about ten minutes, he looked up with a puzzled expression. "Hmmmmmmmmmm."

I refused to take the bait and he went back to the play.

I bypassed another "Hmmmmmmm" and a "Well, really!" but then there was a plaintive, "I find this *very* confusing!"

Succumbing, I sat on the arm of Howard's chair and looked over his shoulder. It was a play in four acts and thirty-seven scenes and the first five pages swept rapidly from a love scene in a garden to the railway station where we saw the girl take a train. The train-boarding scene was followed immediately by a scene *on* the train where she met an ominous stranger. It was all so full of action that there wasn't time for much dialogue, which was just as well; what there was proved strained and painfully elegant.

Occupation: Angel

When the scene changed for the fourth time in as many pages, I started to burst out with: "It's a monstrous joke!" but caught myself in time and left the room on some pretext or other. This time I phoned Buck Crouse. He was in high humor and told me that the play had been written by his laundress, who had acted on the theory that her boss was in a mighty handy little racket which she would like to share. She had proffered the manuscript hand-written on a yellow pad, saying that maybe he wouldn't welcome competition from another writer and, if that were the case, perhaps he wouldn't want to produce it. After reading three pages, Buck was convinced that he didn't. However, she was an unusually faithful laundress and not everyone could do up his navy blue shirts without streaks of the white starch showing. And so Crouse said that indeed his producing calendar was full but, to the delight of the author, he made the gracious gesture of having the manuscript carefully typed in conventional play form and neatly bound in smart blue covers so that it would give a fine, professional appearance. He also had an extra copy made for himself which he planned to use in frightening away unwanted backers.

"Keep him at it another fifteen minutes before you break down and tell him the truth," Crouse chuckled. "Oh, yes, and then tell him the moral is: Don't wake me up in the morning!"

* * *

There is absolutely no wifely fun in rubbing in the laundress story. Howard is always the first one to tell it on himself!

As a team we have been effective but far from infallible. Howard's preoccupation with figures and profit-and-loss percentages plus the ability and patience to understand all the clauses in the contracts are invaluable. And although he would not admit it for the world, I think he has his own divining stick for sensing a dishonestly set-up theatrical deal. I infinitely

A Star in the Kitchen

prefer my share of the teamwork: reading and evaluating the plays. I have no sense of embarrassment about my errors and yet, it's always that tenth play—like the fish that got away—that rankles. And I am well aware that even among the successes, there are some I managed to catch on to only by the skin of my teeth. In fact, *The Skin of Our Teeth* was one I almost lost out on:

We were spending the weekend visiting in Easthampton and tucked in my suitcase was Thornton Wilder's play. The first thing upon arrival, I checked the bulb in my bedside lamp. It was slightly on the skimpy side but I needn't have worried about reading comforts. Our host and hostess had every minute of our visit arranged, and the play had to be put aside for the train ride home.

Even then I had little privacy. I had not counted upon the remorseless gaiety of the weekend playboys who also piled onto the train. As soon as Howard went into the club car for a smoke, an innate sense of gallantry made them try to rescue me from the dilemma of having nothing to do but read! I turned down an offer of a drink in the club car and declined to make a fourth at bridge. Then, to show how happy I was in my solitary state, I beamed brightly and waved the blue-covered script. This seemed to solve the problem until one of the gentlemen untangled himself from a tennis racquet and came over to join me. "Saay! *I'm* interested in the theater," he announced. "You find another lil ol' gold mine like *Life With Father* and I wouldn't mind being an angel, too! How's *this* one—so far?"

In answer I held up the script, with my right forefinger carefully marking page seventeen of Act One.

"Too early to decide, huh?"

"Much."

"How does it start?"

"The maid, Sabina, comes in and says: 'Oh, oh, oh. Six

o'clock and the master not home yet. . . .'"

"Kinda corny, huh? I always think it's sort of obvious when they have to have a butler or a maid answer the phone and give a general exposition . . . and what kind of a name is that for a maid, *Sabina!* Isn't that a Greek name? Sure it is," he continued, answering himself. "The Sabines lived somewhere up in the mountains of Greece . . . used to rape all the women." Then he added archly, "Not a bad custom."

"I thought the Sabines lived in the hills above *Rome,*" I said, in an attempt to stay with the facts, "and I always thought the Romans raped *them,* or at least their women."

Our voices must have carried across the aisle. "Hey!" one of my seatmate's pals called over, "Who's Sabina? And who raped her?"

I could feel the waters of confusion beginning to close over my head; the best I could hope for was to tread water a bit and keep quiet.

"Nobody! It's just a classical reference, you dope! This play —what's it called?" Douglas, my self-appointed companion, looked down at the title page. "Oh, yes, 'The Skin of Our Teeth,'" he read, "'A Fantastic Comedy in Three Acts by Thornton Wilder.' Mmmm. Now, in *Life With Father* the Day family always had reasonable-sounding maids like Bridgit and Margaret; but in this one the maid's called Sabina. I think they ought to change it," he said with resolution. "It's outlandish."

"I don't know," the Pal said reasonably, "you don't want to make it *too* much like *Life With Father*. It ought to have its own flavor."

My lack of response must have bored Douglas, who suggested again that we all go get a drink.

"You go," I urged sweetly. "I simply must get on with this."

He got up and patted my shoulder. "Okay, let's leave her name Sabina," he said with Sunday-night largesse.

A Star in the Kitchen

When Douglas and his pals had migrated to the club car, I started at the beginning again, but the play would not take focus. I stopped trying to see it and tried just reading it as if it were a straight narrative. This didn't help. I attempted to follow the hopes, dreams and tragedies of the Antrobus family from the ice age on to a convention in Atlantic City and back to the Napoleonic War. But nothing really came through to me; I clumped through the play as if it were a muddy bog I had to slosh through in oversized rubber boots. Finally I closed the script; then I closed my eyes for a moment to see if I could kindle a spark of understanding. Howard returned and, seeing my finger in the last page, asked, "How is it? I just had a chat with Doug who says it's another *Life With Father*."

"Douglas is a fool," I commented because it made me feel less of one.

"Maybe he just isn't experienced about plays. Why'd he think it was so much like *Life With Father*?"

"I haven't the slightest idea," I said tartly.

Howard chose to overlook my sharpness.

"It's a comedy?"

I nodded.

"Family life . . . children?"

I had to agree. Then he wanted to know if they had trouble with the children . . . and the maid? When I admitted that they did, he said, "Well, you see! No doubt some of the situations are different, but it sounds to me like basically the same theme as *Life* . . ."

"Don't say it," I threatened. Then I tried to compose myself and give him a brief outline of the story. I failed miserably.

"It's kind of hard to *tell*," I apologized, "because it skips around so. You see, the father invents things, like the wheel and the alphabet; the mother is a sort of symbol of motherhood, the maid Sabina I think of as Lilith and one of the children, Henry, is really Cain; and they all go through things

Occupation: Angel

like the Ice Age and famine and war and somehow they pull through by the skin of their teeth," I finished lamely. "I'm not sure I even know what the play means to say," I continued, "unless it's that somehow we keep progressing in spiral circles; falling down, picking ourselves up and going on; constantly repeating our patterns . . . !"

"It makes no sense to me! And if you admit that *you* don't understand it . . . " Howard picked up the play and made the gesture of cutting off its head.

"Maybe if I thought about it for a couple of days . . . ?"

"What about the audience?" Howard wanted to know. "You going to give them a couple of days between each act?"

The train was pulling into Pennsylvania Station and the occupants of the club car hurried back to their possessions. Douglas greeted us and wanted to know how the play came out—did I like it and should we form a syndicate?

Howard grinned and shook his head: *"Too* much like *Life With Father;* after all, we don't want to go on and on repeating the same patterns, do we?"

No further mention was made of the play for two days and then I admitted that the theme had been niggling around in the back of my mind and I'd been looking for the play to reread it.

"Too late," Howard said firmly, "I've sent it back saying we were not interested."

I thought of Virginia Woolf's *Orlando* and wondered what the first reaction to *that* had been. Still, I was half pleased that the decision for this one was out of my hands.

Sometime later I had a telephone call from Jimmy Byram who told me that he'd seen a rehearsal of the play which had reaffirmed his original faith in it.

"I can't understand *your* not liking it although, God knows, enough other people don't! It's opening in New Haven next week and the producer is still short 80 per cent of the financ-

A Star in the Kitchen

ing."

Only a fool, I reminded myself, never changes her mind. "Okay," I said, "I'm in. Cullman Brothers has turned it down, but I think this one is on *me*. Will *you* tell the producer? Oh, yes, and Johnny, tell him to please not mention it. It's kind of a *surprise* for Howard."

The night the play was opening on Broadway, we were among those present. It had been no problem to persuade Howard to go; we were in our most compulsive theatergoing stage and attended *everything*.

The usher showed us to our seats—fourth row center—and Howard was visibly impressed.

"Damned nice of them to put us so far forward," he observed. "You'd think we'd backed it."

I took a deep breath; the house lights were dimming and it was too late for any discussion. "We have," I whispered and the play began.

* * *

One of my major errors was *Harvey*, which I read and turned down. In fairness to myself (a gesture which I am all too prone to make) I think, under similar circumstances, I would make the same decision again today. And, actually, the margin by which the play moved from doubt to success was narrow.

When I read Mary Chase's play, originally called *The Pooka*, I was charmed with the mad, offbeat whimsy, the appealing story of an amiable drunk and his constant companion, an Irish fairy spirit who took the form of a six-foot white rabbit.

Perhaps my lifelong devotion to *Alice in Wonderland* made me resent this Pooka as an oversized interloper. In any event, it was my contention that one couldn't get away with having the giant rabbit wandering around the stage for three acts. I didn't know the producer, Mr. Pemberton, but I tentatively

Occupation: Angel

posed the problem to the intermediary who had sent me the script. He was properly shocked and reminded me that the play was about a pooka and I couldn't very well suggest leaving him out. He was sure that Mr. Pemberton would take a very dim view of such a suggestion. I said in that case to count us out. At the tryout in Boston, the director, Antoinette Perry, viewed the performance impartially—it just wasn't very funny; the man in a rabbit suit left something to be desired. She proposed that they try it without the pooka. Sure enough, Mr. Pemberton was *not* agreeable to the suggestion and reminded her that not only was the play (now renamed *Harvey*) called after the character but also that he had just paid a bill for $600 for the rabbit suit—the most expensive costume in the whole play. Finally, at the end of the week, the director had her way and, just as a trial run, the matinee performance was played without Harvey's physical presence. It got twice the number of laughs. The director won, the play was a hit, and Mr. Pemberton even worked out a system to soothe his sense of outrage at the loss of $600. The costume was used after all: the rabbit came on for the third curtain call!

* * *

During the height of the season, we go to the theater three and sometimes four times a week. We have an assortment of reasons why we must go to the opening: either we have backed the show or we have turned it down, and in either case we couldn't possibly wait for the second night to see if we were right. Or, we never had a chance at the play and we must see if we should congratulate ourselves or commiserate.

Dates for opening nights are constantly subject to change and some hostesses, quite reasonably, resent our calling at a late date and saying that the dinner invitation we had accepted three weeks earlier now falls on a conflicting date. . . .

After the show, the general rule is for little groups to drop

A Star in the Kitchen

in at Sardi's or Twenty-one for a drink and some supper. There the news filters in quickly and the message, "Mixed notices," or "*The Times* gives it a rave," bounds from table to table.

Occasionally a show, usually a musical, meets with such outstanding success out of town that it comes in with all the assurance of an acknowledged hit. And then the producer may feel lavish enough to take over the roof garden of a hotel and make a big splash. Such events are rare, but they have all the elements of glitter and glamour one usually associates with the stage. A party after a great hit is a christening, a birthday, Christmas, New Year's Eve and a wedding all rolled into one. By the time the star of the show receives in her dressing room, admires her flowers, reads her telegrams, takes off her stage makeup, puts on a fresh face and changes from her costume to an evening dress and arrives at the party, a good hour and a half may have elapsed. The party might have started earlier but her arrival is the real signal that the party has officially begun.

Unfortunately for Howard and me, although the first two hours may find us in fine fettle, this is the moment when visions of a new day that starts at eight for Howard and at nine for me begin to intrude. The knowledge that for most of the others, day dawns with the noontime whistle is small consolation to us. It often is difficult to live in two worlds, not wholly belonging to either of them.

It is rare that anyone feels so fully confident of a show's success that an elaborate party is planned for it. Nothing is so awkward as a gathering where everyone knows that the play is going to be panned by the critics and probably deserves it; it is hard on everyone concerned. However, we like to take a chance on occasional home parties with no more than forty or fifty people dropping in. Somehow they are always the warmest and the gayest; and the conversation may have more sparkle and polish than the lines that we have just heard on

Occupation: Angel

the stage.

At one such party, Elliott Nugent and James Thurber outstayed all the others. At two-thirty in the morning, Howard quite sensibly went off to bed; I stayed up to talk and to laugh. It was one of those magic times when all our thoughts seemed to come out well shaped and funny before we knew we intended to say anything at all. It was a moment to cherish, to cup in one's hands and hold; but, like any other magic, it cannot be held. It was intense for the moment; the next day nothing remained but a warm, intangible glow.

Far from disrupting the tenor of the household, the parties seem to give a little extra fillip. I not only never lost any of my help as a result of the suppers, I even gained an extra two years' time with a cook. It was more than just the party; I think I owed it all to Ingrid Bergman.

We had a marvelous Swedish cook, talented, agreeable and a born party planner. She used to say that anyone could plan a formal eight o'clock dinner party but a midnight supper required care, imagination and delicacy. She shared my antipathy for the dull, baked ham and cold turkey that look as if they had been created solely to display on TV—all glaze and no taste. We agreed that food at that hour should be hot and very light. She could bake a huge pastry shell shaped like a boat but so fragile and flaky that one felt that at any moment it might take to the air. This she would fill with a mixture of half chicken, half sweetbreads, with flecks of pungent truffles in a rich, smooth sauce. There were always eggs; sometimes a huge platter of them scrambled, or an endless supply of fresh, individual omelettes surrounded by dishes from which one might choose the accompaniment: crisp bacon, chicken livers, or a delectable concoction of kidneys and mushrooms in a madeira wine sauce. This dish was part of the supper at the party we had after the opening of Ingrid Bergman in *Joan of Lorraine*.

A Star in the Kitchen

All of the food seemed especially appetizing to me that night, possibly because just the week before Cook had told me that when we moved to the country that spring we must go without her. The time had come for her to retire. She had it all figured out: we had been generous, she had been frugal; she would get herself a cat for company and take a small apartment in Long Island near her friends. It was a well-thought-out plan, and I did not have the heart to try to dissuade her. Meanwhile, we would have a few months more in which to enjoy our blessing.

By the time Ingrid Bergman arrived at the party, vast inroads had been made on the food and there was barely a dab left of the kidney dish. She sniffed it appreciatively and tasted what there was. Then her eyes lit up with pleasure and she declared that she had not tasted anything like it since her girlhood in Sweden. We went back into the kitchen together to see if Cook had any left. There was a generous portion left in the pot on the back of the stove and Ingrid scooped it up delightedly. She complimented the cook and suddenly the kitchen was alive with the lilting singsong of the Swedish language. As we left the kitchen, I looked back at Cook; her eyes were shining with soft pleasure, her cheeks were flushed and she looked almost girlish.

The next morning, hard on the heels of my breakfast tray, came Cook.

"Last night I could not sleep," she announced.

"Oh, I *am* sorry, did we make so much noise?"

"I could not sleep because I am *thinking*," she said, "I am thinking: 'Have I spent nearly fifty years perfecting myself to cook for a cat?' And when my friends have their Thursday afternoon off and come in to see me, shall I be able to tell them how Ingrid Bergman looked and what she wore, and what she said about my cooking? Or shall I tell them that my cat has fur balls—and sound like a stupid old woman?

Occupation: Angel

"And *you*, Madame, might have to hire a *German* cook." (This was tantamount to wishing me leprosy!) "No," she said with an air of finality, "it is unsuitable, all around. And if you still want me, I'll put off retiring for another year or two."

I nearly upset the breakfast tray when I threw my arms around her. We had a warm, emotional embrace and just missed weeping. I am inclined to be a little unstable anyhow with a hangover.

11

Readings . . . Auditions

As I have said, I prefer to read a play script by myself, but current musicals present an entirely different problem. A quiet, solitary reading offers little clue to their real potential. For an amateur, the musicals of today are almost impossible to judge without an audition—and *extremely* difficult to evaluate even then.

Musicals have come a long way since my childhood, when they were always referred to as musical comedies. The ones I remember could be best described as the very sketchiest of boy-meets-woos-wins-girl stories which were interrupted at regular intervals with painfully appropriate songs and dances. For example: He and she have just met at a ball. He is unaware of the fact that she has long and secretly worshipped him from

Occupation: Angel

afar and she is unaware that he is about to become betrothed to a very spoiled, snippy girl. They waltz onto a balcony overlooking a flower garden and suddenly, having taken a good look, he says: "In the moonlight your face is like a flower." This botanical phenomenon having been disposed of, he might go right into the number "Someday I'll Find My Love in a Garden." And, after two verses, she'd figure it was her turn and she'd upstage him two paces while she confided to the audience that for her there was no "when" she already "had." While she sang, he'd take care of a lot of little chores of his own; he might count the house, check the degree of polish on his fingernails or even pick a bit of lint off his lapel. That accomplished, they'd join hands and sing to each other. Meanwhile a dozen young women, artfully done up to represent the different flowers in a garden, would drift out one by one and perform a flower dance. There was a better than fifty-fifty chance that a dozen young men wearing the exact facsimile of his costume would join the dance and might even wind up each picking a girl flower for himself.

This kept everybody pretty busy so that fifteen minutes, and a lot of flowers later, the story line was right where it started and nobody knew anything more about the characters, but that wasn't expected. And everyone knew how it was going to come out—the fun was in seeing it done.

Successful producers such as Cohan or the Shubert Brothers turned out a succession of such musicals on budgets as low as $15,000 and rarely in excess of $25,000. And, at that, only a small percentage of the sum had to be available in advance; more than half of the initial expenses were held over and paid out of the show's eventual profits. It was well within reason for these shows to have paid off and begun to show a profit within one month. It is true that shows in general did not enjoy very long runs, but financially it was not necessary. Producers were not obliged to put up much money to start

Readings . . . Auditions

with, they had no backers with whom to share the profits and, as they often owned the theater in which the show ran, even the rental price went into the same pocket.

A musical that ran for five months was considered a big hit. A short run was the rule rather than the exception. Even big extravaganzas like *The Follies* or *The Scandals* were completely changed each year and, to prove that these shows were new, the current year's date was always attached to their names.

Then, slowly, the American musical began to come of age: No longer was it necessary to say "musical comedy"; a musical could tell a real story, even a sad one; and the songs and dances ceased to be an interruption but became a natural extension, clarifying the characters and their motives and developing the story line. Concurrently, costs began to rise, not only because of generally increased prices, but also because stricter theatrical rules were applied. To launch a show, the producer had to have the whole budget in hand instead of a third or a quarter of it. Costumes had to be paid for in advance; and posted bonds guaranteeing round-trip travel expenses and at least two weeks' salary for the cast brought an end to the old stories of stranded theatrical troupes.

The expenses involved in financing quadrupled, and producers faced the fact that each musical represented an outlay of from $60,000 to $100,000. The prices of theater tickets had risen, but not in proportion, and there were no bigger theaters to help amortize the heavier outlays.

Outside money was needed from backers and Broadway evolved the audition as a way of getting it. Sometimes called "rustling for dough," sometimes "singing for next season's supper," the audition is an informal recital of the story, the music, and the lyrics of a projected musical show.

The custom of enlisting the partnership of backers or angels has continued; in fact the need for them has grown greater

Occupation: Angel

as the budgets which sent producers scurrying out for help have once again quadrupled in the past twenty-five years. The one bright spot is that the life of a good show has grown appreciably longer, offering the promise of greater eventual returns to backer and producer alike.

An audition may be given at any hour from matinee time on into the night. It may be given in an empty theater (a rarity), a business office, a musician's studio, a hotel suite or a private home.

George Abbott has the iciest, and possibly the most direct, approach to the subject. He generally hires a music practice room with nothing but a few rows of wooden chairs and a grand piano. He invites one or two music publishers and probably not more than half a dozen backers who are almost professional in their interest in the theater; and because he thinks of this as a business proposition and not a social gathering, he is likely to set the audition for three or four o'clock in the afternoon. Refreshments are conspicuously absent. Briefly, and rather badly, he outlines the plot and tells who, so far, has been engaged for the choreography, costumes, scenery and casting. Then he introduces the pianist, who might also be the composer, and two or three singers who take it from there. Everything is handled as crisply as a prosperous Wall Street syndicate outlining a new and much-sought-after security.

Since this technique is practically a reaffirmation of Mr. Abbott's character, it serves to remind backers that he never has been known needlessly to waste five cents of his own or of a backer's money.

However, the majority of producers who give auditions seem to feel the need of a little more atmosphere and comfort. The studio atmosphere is useful for creating the impression that everyone is in on the very birth of genius; on the other hand, a really elaborate setting, even though borrowed for the occasion, casts an aura of success already achieved.

Readings . . . Auditions

There are those who say somewhat bitterly that Mr. Abbott carries his financial temperance to extremes. Even his most ardent admirers, including Howard and me, have to admit that he is compulsively economical. And for one brief hour or so I am afraid we phrased it less politely.

We had invested in a musical George was producing called *Beat the Band* and word had crept down from Boston, where it was trying out, that it was in trouble. George telephoned and asked us if we would go up to see the show and talk things over with him.

"When Abbott telephones from *Boston*," I said to Howard, "things must be very bad—maybe there's nothing worth saving."

At any rate, we took the train up to Boston, arriving just in time for dinner and the show. I carried a pad and pencil with me and made notes of some of the things that bothered me; I used up a good deal of the pad.

After the curtain, George rushed over to us and announced that he couldn't talk then. I held up my pad of notes. Well, he had to get right to work with some of the cast and it had to be done immediately. But he wanted to hear our reactions . . . in the morning. Howard reminded him that we were taking the ten o'clock train back to New York. Good! We'd talk it over at breakfast: eighty-forty-five *sharp* in the Coffee Shop, the counter because the service was faster.

"And cheaper," Howard muttered. "I really think," he complained to me, "that when we take the trouble to come all the way up, and stay at a hotel overnight. . . ."

"There, now," I soothed, "you know what George is like. No doubt there's some union problem about the use of the stage after midnight. It might involve extra charges. And if these suggestions are really valid, they'll be just as good in the morning."

A counter may not be the ideal setting for a play conference

Occupation: Angel

but it served adequately. We barely noticed the thick mugs and the coffee that slopped over. We saw pretty much eye-to-eye on the problems and their possible solution. Changes would be made, the show would stay out an extra week and hopefully open with more strength in New York.

"Well," Abbott said, suddenly solicitous, "I mustn't make you late for your train—have to get back to the theater myself." He motioned to the counterman. "Separate checks," he said, and shook hands goodbye.

I think we were past Providence before Howard cooled off.

Abbott stories are a tradition in the theater, but not as many people seem to know that Alfred de Liagre possesses a similar idiosyncrasy. Perhaps it is because personally and socially Delly is a generous, gracious host. It is only when a sum has to be listed as an expense against his production or the running cost of a show that money assumes a new importance.

It was shortly after the opening of *The Voice of the Turtle;* we were picking Delly up at the theater and all going on to a party together. The play had received wonderful notices, the box office could scarcely fill the demands for tickets and although none of us could guess at that point that the show eventually would yield a return of 1,800 per cent on our investment, we were all feeling confident, secure and, I would have thought, even lavish. Maggie Sullavan stuck her head out of her dressing room, greeted us and reminded Delly that she was still waiting for some carpeting.

"I know. I know all about it—I'm *working* on it," he answered. "And just as soon as I can find it, you'll get it."

"Actresses," he continued to us on the way out. "Honestly, they are so *impatient!* When Maggie makes a quick change, she has to run across backstage and her heels clatter on the bare wood floors. Just last week we agreed that we need a narrow, very long strip of carpeting. . . ."

As he was talking, I was mentally adjusting the problem. It

Readings . . . Auditions

would be amusing, I thought, to call up Sloane's and order a forty-foot runner sent right over—color and quality, no object; it's just for *back*stage. . . .

". . . about thirty-five feet long, and it doesn't need to be more than two and a half or three feet wide, although it *could* be more. Altogether, an odd shape," he continued, "and I'll bet you I've tried half a dozen places! One can't immediately pick up a size like that at Cain's Warehouse—or any other secondhand prop house."

In a world of unreality and extravagant gestures, I find this touch of New England frugality eminently reassuring and satisfying.

* * *

Invitations to auditions sometimes arrive by telegram, sometimes by phone. One may receive a formal, engraved invitation, a personal note, a form letter or a printed card. It may come from someone you've never met. The less standing the producer has, the more frantic his approach so that sometime you'll even find an ad in the daily paper:

> BROADWAY MUSICAL SHOW
> OPENING SOON
> Offers Private Individuals
> Opportunity for Investment
> References Exchanged

If the function of an audition is to give the backers a chance to "see" the show in advance, then it usually must be considered as a failure.

There are about six vital, component parts to a musical. They are (not necessarily in the order of their importance):

(1) The book
(2) The songs (music and lyrics)
(3) Casting

Occupation: Angel

(4) Visual aspects: costumes, scenery, staging and lighting
(5) Choreography
(6) Direction

And, in some cases, (5) and (6) may be combined under one person, a choreographer who also directs.

At an audition, the book generally is disposed of in the briefest sort of outline—a mere skeleton upon which to hang the songs; the casting rarely is settled, nor are the scenery, costumes, direction and choreography.

What the backer hears musically is not even an accurate one-sixth because there is a very good chance that several of the songs will assume entirely new proportions when they have been fully orchestrated and one or two of them may be dropped out of town and new ones substituted.

"I Hear Music . . .," the hit song of *Call Me Madam*, was a last-minute inspiration of Irving Berlin's, written practically overnight to fix up a weak spot when the show was trying out on the road.

The amateur investor who bases his decision on an audition is, whether he knows it or not, actually taking 75 per cent on faith, good will and his confidence in the talent, taste and experience of the producer.

Usually, the potential backer arrives to find something that looks like that sensible but seemingly impossible arrangement: a seated cocktail party. Furniture has been pushed around and extra chairs brought in to form a large semicircle around a grand piano. Most of the other guests may be strangers to you, although a few of them are the names you've seen when you signed partnership agreements for other shows.

With a little practice one often can figure out many of the people, even before being introduced.

At one audition there was a small blonde who anticipated the humor by a split second and led the response with what appeared to be completely uncontrollable gusts of laughter.

Readings . . . Auditions

Not surprisingly, she was the wife of one of the authors. She had attended every audition and knew the book and songs by rote. If she was sincere, she had a rare facility for appreciation that could bear up even under repetition. She must make a wonderful wife.

The homely, expensive-looking young woman who had three people dancing attendance and pressing highballs upon her was a much-feared and courted heiress who has been known to insist upon putting up a quarter of the financing and then capriciously backing out at the last minute.

The knowledgeable host will make the fewest possible introductions, leaving it to the individuals to assume that all the unknowns are fascinating and important people.

Customarily, the composer plays his own music and a couple of young actors and actresses will sing several different roles. Usually they are hired for the occasion and have nothing to do with the eventual casting but merely represent a reasonable facsimile of the characters to be portrayed. Sometimes they are students earning a little extra money, sometimes they hope to be cast in a lesser part in the show or act as understudies.

According to Equity rules, they must be paid at least the minimum fee the first time they are engaged and twice that for all subsequent performances.

The rules also state that they may read their lines or sing but may contribute no "movement" which might be interpreted as "acting." There is also a time clause which stipulates that, regardless of breaks or intermissions, not more than four consecutive hours of the actor's time may be used.

This performance is supposed to offer the potential backers an hour of lighthearted fun and the conviction that they are seeing a hit in the making. It is a period of nervous strain for the authors and producer, and it would seem to be an unlikely moment for a guest to conduct private business of his own. Yet, invariably, someone will announce loudly that he is expecting

Occupation: Angel

a call—an important one, he will explain as if the importance precluded any need for apology. When the call comes through, he may decline the offer to take it in a back bedroom, but will elect to talk within sight and sound of the assembled group. One hand presses the phone firmly to his ear, the other hand covers the mouthpiece; his conversation is cryptic, staccato and his eyes wear an expression of hooded secrecy. At a given point, he may signal wildly for a pencil and paper; then, with hunched shoulder holding the phone to his ear, he will scribble earnestly. All eyes are drawn to him like a magnet: Is he a statesman getting an ultimatum from Khrushchev—or an agent receiving word from Rosalind Russell that she will agree to star in the show? As he returns to his seat, he may exchange a significant look with someone: raised eyebrows, widened eyes and a half nod. Speculation is rife among us; we do not know if war is about to be declared or a major casting arranged. In either case, the performers are in for a sharp lapse of audience attention.

Repetition of such occurrences has given me a jaundiced eye and I always am convinced that the gentleman's wife has phoned asking him to bring home two extra lamb chops because his sister-in-law is coming to dinner. But the histrionics are wonderful and usually much more professional than those of the young singers who have been engaged for the performance, or of the author or producer who outlines the story.

Way back in the spring of 1947, Howard and I attended an audition in a private house. It was a beautiful evening and the drawing room opened onto a charming little city garden skillfully planted, lighted and equipped for the occasion with a small but bountiful bar and a solicitous butler. There was a rather unusual air of social correctness combined with artistic informality. The audition seemed to me to be a great success—the plot was funny and the music and lyrics a dream. Everyone connected with the show was talented and I had confidence in

Readings . . . Auditions

the producers.

The very next morning, we subscribed to four shares at $3,600 each and congratulated ourselves on having got to the first audition.

Several months later, when we returned from the country, we found out that we had been the sole substantial backers, not only of that first evening, but of the endless, exhausting auditions that had followed. The performances had become a great social success, a rather chic form of entertainment much appreciated by summer bachelors and bored fugitives from Southampton. But, by and large, the audiences continued to keep their money in stocks and bonds.

Eventually (over a period of six months) the money was accumulated in dribs and drabs, but only after the units had been broken into halves and quarters for small investors.

Shortly before the show left for its out-of-town tryout, Howard and I attended not a run-through, which is the equivalent of a complete rehearsal without sets or costumes, but a slug-through, which is an earlier, rougher period of rehearsal with frequent interruptions and repetition. For fully three quarters of an hour we watched while the heroine banged a bare wooden table with a pewter tankard and sang: "I Hate Men."

"Now don't bother to really sing it—save your voice," Cole Porter urged. "Let's just run through the 'business' of that again."

The fourth time they went through it, I began to squirm. Howard and I exchanged a look solid with the sense of potential disaster. If the leading lady couldn't even bang a table to suit the director! I had twinges of conscience about the friends whom I had persuaded to invest in the show. At that moment I would have sold our share of "Kiss Me, Kate" for fifty cents on the dollar. And if I had, I would have been a fool, because the show was an enormous success; several critics

Occupation: Angel

commented upon the spontaneous impact of the tankard-banging scene; and eventually we got back nearly five to one on our money. (Despite a popular misconception, musicals almost never pay off as well as straight plays.)

Of all aspects of play judging, for me musicals have proven the chanciest and the most difficult to evaluate.

Since musicals have come of age, it is almost impossible to come to any decision on the book alone—unless it is patently poor. The book, music and lyrics go together and must be judged that way. This makes it hard on the producers, many of whom find the role of barker essentially distasteful. Yet even though they may avoid the carnival trimmings, most of them are trapped into holding these auditions as their only means of raising the money needed. Not more than one musical in twenty comes to fruition without the need of fund-raising auditions.

Some years ago, a couple of lads who had attended the same college banded together for their first effort at producing. They had a lot of bright ideas for skits and a lovely duplex apartment in which to give auditions—only they couldn't find anyone to come. This seemed such a pity that a couple of relatives and close friends (graduates of the same college) rushed in to help.

Somehow, in the ensuing melee, the idea of investing in a show became confused with devotion to the alma mater, which appeared to be in peril unless everyone rallied round and backed the young graduates. Everyone was too busy giving school yells to remember to give the business details, which actually were quite attractive. All available classmates were telephoned and, when they ran out of one year, they went back a whole generation.

After such a telephone call, one old grad hung up the phone in a state of utter confusion: "Damn funny way of raising funds

Readings . . . Auditions

for a new building," he kept muttering.

Another producer had in preparation a musical whose plot revolved around a local historic area. The owners of all the large, surrounding estates were telephoned and told that through sheer civic pride they should back the project. It even was implied that a successful production would bring additional renown to the area and increase property values!

One producer gave so many auditions that potential backers going in would meet others coming out. It was like a neighborhood movie house, and if you missed the first act you could stay on and catch it in the second performance.

I was aware of this problem and in a sympathetic mood when I had a telephone call late one afternoon from Herman Levin. He identified himself on the telephone as the co-producer of *Call Me Mister*, in which we had invested.

"Of course, Mr. Levin, I remember you very well," I said, and we chatted about how pleased we were with the success of the revue.

Then he told me that he was doing a new musical and wanted to talk to me about it.

"You mean you want me to come to an audition?"

"Eventually. But first I want to talk to you about it. Could I come over, *now*, to see you?" There was a note of urgency in his voice.

When my husband came home, Mr. Levin and I were having a drink and still talking generalities about the theater. I found him exceedingly bright, funny and, as we agreed on most points, I thought him sound. He outlined the plot and his plans for his new show. When Howard joined us, Mr. Levin got down to the subject of finances: the musical was to be capitalized for $200,000—he'd already given several auditions but he was still short of money.

"Just how short?" Howard wanted to know.

"A hundred and ninety-five thousand short," Levin admitted

Occupation: Angel

with charming candor. Then he went on to say that the real hump was the next $100,000; people were reluctant to go in until a large part of the financing was accomplished. The book, he assured us, was in excellent shape, the music and lyrics absolutely first rate and funny. He'd be happy to have a private audition for us in our own home at any time.

Howard explained that $100,000 was too steep for us. He looked across at me with raised eyebrows. I nodded and he added: "But maybe we could interest a few friends."

"I can't *promise* anything, I haven't even heard it yet myself but I'll take the first step on faith and invite a few friends over. If you'd care to give the audition here one evening next week, we'll get a chance to judge, and perhaps we can help you," I said.

When Herman Levin and a few of his associates arrived the following week, there were no more than twelve or fourteen people having coffee after dinner and, as he pointed out later, there wasn't a familiar Broadway face in the crowd. The author and composer looked at the producer accusingly. This didn't look like ready money.

There was nothing to do, they said afterward, but go through with the audition with as good a grace as possible. And, at least, the audience proved friendly and receptive.

We were more than merely receptive, we were completely captivated.

The producer outlined the story for us. Some people are born storytellers, some are born story-messer-uppers. Herman Levin belongs to the first group; he can read three names and numbers out of the telephone book and I am rocked with laughter.

A young woman and a young man had done the music and lyrics and they modestly said they'd try to sing for us themselves; and we must use our imaginations to see the scenes as they would be with proper casting. None of us had any idea that they were skilled professionals. And, unfortunately, no

Readings . . . Auditions

one else performed in the show one half as well as Betty Comden and Adolph Green.

At the end of the performance, I scribbled down our initials and $20,000 with a question mark. I showed this notation to Howard, who nodded. I had a quick word with the other guests who were interested and scratched down the amounts they suggested. It wasn't adding up properly and I was afraid that the producer, who appeared rather uncomfortable, was embarrassed at being present when I talked finances.

I suggested that I'd better call him in the morning, as I was having a little problem.

He says that he barely restrained himself from saying: "You and me both!"

But the next morning when I telephoned Herman Levin, I explained that I had it all worked out. I had planned to keep $20,000 for ourselves but, by cutting that down and paring some of the others a trifle, I had gotten the pledges to come out at an even $100,000.

There was a sudden and almost audible silence on the other end of the phone.

"Mr. Levin! Mr. Levin! Are you there?" I kept repeating.

Later Herman said that the rumor that he fell into a dead faint was entirely unfounded—he merely dropped the telephone and bumped his head when he tried to pick it up.

It would be nice to reminisce about the enormous hit the show turned out to be and how much money all of our friends made on it. Alas, it was one of the quickest failures on record; it dropped the full capitalization and closed before it ever reached Broadway.

We went down to Philadelphia to face the bad news before the show closed. We sat there in discomfort while the jokes never quite came off, the plot stumbled all over itself and the songs fell flat.

"I just wish," I said glumly, "that we had that *original* per-

Occupation: Angel

formance back." The material, in truth, was best suited to precisely the performance we had seen: two charming, ingratiating and talented young performers entertaining an intimate group of people. Some of the material eventually became part of an act which Comden and Green used in night clubs for several years afterward.

The ebullient Mr. Levin bounded up with fresh plans: a musical version of *Gentlemen Prefer Blondes*. I was cool to the idea, but not because of the previous failure; I simply couldn't vision it. By this time we had grown to know Herman well and were very fond of him. We kept our minds open and went to an audition. It didn't change my opinion.

"I'm sorry," Herman said earnestly. "I feel very sure of this and I hoped you could make up your loss."

I declined.

"Don't just stay out of it and do nothing!" Howard commanded. "You're a good friend of his; sit down and have a talk with him and persuade him not to do it. It's sure to be a disaster."

"I can't do that!" I was equally firm. "He knows his own mind. I doubt that I could change it if I tried—and I won't try. After all, maybe he's right and I'm wrong."

That proved to be the only aspect in which I *was* right. The show made Herman Levin a small fortune.

* * *

On the evening we had tickets to see *The Diary of Anne Frank*, which we had missed opening night, Howard telephoned and announced that he would have to go to a business dinner.

"Tell you what," he suggested, "I could call Herman Levin and ask him if he'd take you."

Half an hour later, Howard reported that Herman would be *delighted*—nothing would please him more than to take

Readings . . . Auditions

me to the theater. And, besides, Herman had some business he wanted to talk over with me. Since I automatically discount at least 20 per cent of Howard's version of other people's enthusiasm for me, I figured that what Herman *really* had said was: "Oh, sure. I don't mind; I haven't seen the show either."

In any case, we had a fascinating evening at the theater and, to our surprise, found the play profoundly touching without being depressing. We walked homeward as we talked about it. Herman suggested that although it might seem like a highly irregular way for a producer who was out with another man's wife to behave, he'd like to go to Schrafft's for a hot fudge sundae.

"A Luxuria ice cream cake with hot fudge and almonds," I bargained.

"Sold!"

"And the business?" I asked after we had been served.

"I've got an idea for another musical. What would you say to the idea of making a musical out of Shaw's *Pygmalion?*"

I thought this over very carefully while I ate two spoonsful of the ice cream cake. I savored the rich, hot fudge sauce against the cold ice cream and the ever so lightly salted, crunchy almonds. The Luxuria ice cream cake was one of the few remembered gastronomical glories of childhood and adolescence that still tasted the same.

It occurred to me that I must have been in the same pre-adolescent stage when I first encountered and thoroughly enjoyed *Pygmalion*. When I read the play again a dozen years later, I realized with surprise and pleasure that I had missed very little the first time. Like the Luxuria, it had a staying quality.

"I like the idea very much indeed!" I said.

"Of course, it's been done rather often," Herman pointed out.

"But not as a musical!"

Occupation: Angel

Herman looked less pleased than I had expected.

"Don't tell me that you just like to argue!" I said. "And that when I agree with you, you're going to take the other side! Aren't you *glad* that I like the idea?"

"Yes and no. If you hadn't liked it—there'd be nothing to discuss. This, as I vision it, would be an expensive production. Maybe $400,000. . . ."

I gave an appreciative half-whistle. In those days, that was a new high.

"Bill Paley of Columbia Broadcasting System is crazy about the idea and wants to finance it. *All* of it!"

Herman went on to explain that if Mr. Paley had offered to put up three-quarters or even four-fifths of the money, he would have been delighted—but putting up all of it meant squeezing out us and other loyal backers. Herman was concerned that he couldn't hold out even $25,000 for us—and he had to give his answer to Mr. Paley in a couple of days.

"But of course you must say yes and go right ahead. Why it's wonderful for you; no auditions, no scrounging around for money, all of it there in one glorious, big, lump sum."

"And you really don't mind?"

"Of course I don't mind," I insisted gallantly. "It's too good an opportunity for you to risk messing it up, just to cut us in. Besides, it's just *one* show—and a very expensive one. You'll do another one next year . . . and the next year . . . and the next year."

Once again, I was wrong. The producer of *My Fair Lady* was too busy with the original company, the first road company, the second road company, the London production, cast replacements, negotiations for a movie deal and productions all over the world. Year after year for six years the money rolled in and any further producing plans of his lay dormant.

Howard takes a very dim view of my occasional suggestion that we stop in at Schrafft's while I have an ice cream cake.

Readings . . . Auditions

"Do you know what that ice cream cake that Herman Levin bought you cost us? Well, figure it out for yourself: *My Fair Lady* was the longest running musical on Broadway and, counting the productions all over the world, it must have grossed well over $50,000,000; the motion picture rights were sold to Warner Brothers for $5,500,000 down plus a percentage of eventual profits after the movie is made, and that's the highest amount ever paid by a film company for *any* literary or stage property.

"You and your cavalier gestures! I figure that if we had had $25,000 in the show . . ."

Sometimes I think that Howard has a very morbid turn of mind.

12

The F.B.I. and I

One of the paths the theater took me down led straight to a personal investigation by the F.B.I. It was wartime and I was suspect as a spy! Certainly this, above all else, would have justified Grandma's worst fears; I was the first member of the family to be suspected of being a traitor.

Chronologically, the incident started long before we entered World War II. A Swedish-born friend telephoned to ask a favor. Would I see one of his countrymen who was coming to America for the first time? The gentleman wanted to make theatrical contacts with a view to buying American plays for Scandinavian production.

I flipped over my calendar quickly: the daytimes were crowded, but we were giving a dinner party the following eve-

The F.B.I. and I

ning for a somewhat older group of friends. It would be an easy matter to slip in an extra man and for some reason I assumed that he would fit into the age group. The Swedish gentleman and I were introduced over the telephone and Mr. Schmidt was pleased to accept the invitation.

The next morning I had an invitation for an impromptu cocktail party: there would be half a dozen of the Artists and Writers crowd whom I had not seen for years.

I said that I had an early dinner party on my hands, but I was easily persuaded that I could manage to get home in plenty of time. I took the precaution of wearing the dress I had planned on for the evening; I checked the wine, admired the flowers, cast a final approving look at the table and started to put out the place cards. For a moment I hesitated over the one lettered LARS SCHMIDT. He would know no one, I wasn't sure how good his English would be and he might be hard to talk to. He would be my responsibility, I decided, and propped up the card on my right.

The old friends at the cocktail party proved even more beguiling than I had remembered and time took on its own tempo.

That evening, for the first time in my life, I was a half hour late for my own party and most of the guests, with the promptness of the older generation, had already arrived. They were grouped in little cliques, chatting and sipping their drinks. Someone put a drink in my hand and I gulped it nervously as I began to make the rounds. Everyone's name came easily to my lips and I remembered which ones I thought might have a natural affinity for each other. The party seemed to be going well and I drew a sigh of relief that turned into a slight hic. I looked with consternation at my empty cocktail glass, remembering too late that I'd already had two—my usual quota—and that a third drink was likely to snap back at me. At the same moment I realized that I had another problem: there was a very blonde, tall, thin young man standing at the far end of the room and I had not the slightest idea who he was.

I gazed across the room despairingly at my husband. He interpreted my look as an apology and he returned a look that was half censorious, half forgiving. Two other guests arrived and needed greeting. But where was my elderly Swedish gentleman?

First things first, I decided, and approached the strange young man with a noncommittal: "How nice that you could come!"

His heels came together smartly and he bent over and kissed my hand.

"You are very kind." His English was precise, clipped and with a faint accent.

"Mr. Schmidt!" I exclaimed with relief. "Forgive me. I was confused."

Mr. Schmidt made a deprecating gesture. "Not *very* confused; I should guess only about one Martini's worth," he said softly.

"You mean it shows?" I was horrified.

"Not to anyone else; I'm intuitive," he said.

As it turned out, he was, too. We developed a fine friendship and I introduced Lars Schmidt to everyone I knew whom I thought could be useful. We spent long hours talking about the theater in general and specifically about plays that might do well in Sweden. His interest was twofold: to find old American plays which had never been produced in Scandinavia and to be first in line to bid on the current ones as they came along. In the latter aspect, I could be especially helpful and I offered to make suggestions to whomever he finally chose as his American representative.

"I wish *you'd* take on the job," Lars suggested. "You see all the plays opening night anyway, and you read a majority of

The F.B.I. and I

them first, which would be a distinct advantage. We could have a bid in before the play even opened."

I explained that I knew my limitations and, while I might have a good record for picking plays for an American audience, I didn't know anything about the Swedish people, their taste, their reactions to plays.

Lars continued his search for the right agent, but war clouds were piling up heavily over Sweden and abruptly he was recalled home for military service.

"Are you sure you won't change your mind?" he urged.

I suppose no suggestion seems as impossible the second time you turn it over in your mind. In any case, I agreed to take a try on a partnership arrangement. As a parting gesture, he armed me with a list of twenty or thirty English and American plays that had been produced in Sweden, their notices (thoughtfully translated for me) and a record of their financial outcome. I was familiar with most of the plays and it was fascinating to see what the Swedish reaction had been and finally, as I studied the list even more closely, to see a kind of pattern evolve.

Some of the American plays on the list had been produced in London as well as in Sweden and I soon came to the surprising conclusion that the Swedish sense of humor was much closer to ours than was the British.

I was intrigued with this international aspect of the theater.

Shortly after Lars had returned to Sweden, I had a long-distance telephone call.

"Sweden is calling me," I announced breathlessly to my husband.

Twenty years ago the telephone did not bridge the seas as casually and efficiently as it does today. As a matter of fact, I never had spoken to Europe. But Howard, who often had conducted international business, seemed to find the idea of the call less world-shaking than I did.

Occupation: Angel

Finally, after a series of strange cacophonies, a voice came through.

"Hello, Marguerite? It's Lars! Your business partner," he added, inferring that there might be a dozen other Larses who would be telephoning me from Sweden.

"Yes, I know," I said. "How are you?"

"I'm fine, I . . ." The rest of the sentence was lost in a roar of meaningless sound. Then I caught: ". . . business expense. And how *is* business?"

I tried to tell him but on every fifth beat a wave seemed to break over the telephone wires, drowning out the following word or syllable.

So we stuck to exchanging less pertinent and vital information, such as what time it was in each country and how the weather was. Despite well-known, available statistics on the subject, it came as fascinating news to me that it was five hours later and much colder in Sweden.

After I hung up, I turned to my husband: "My, that was fun."

"I hope it was," he said drily, "as I daresay you've just used up the first month's profits."

From then on, to show how prudent I could be, I made a point of answering Lars's wordy, expansive cables with the most terse, abbreviated ones I could concoct. That is what eventually landed me in trouble.

I sent weekly, comprehensive letters airmail, but Lars would hear about a play I had not mentioned and cable me asking what I thought of it. It was useless to ask Lars to wait for the letters; he was fascinated with immediate action.

We still were months off from our own entrance into the war, but there was a kind of anti-German hysteria pervading the country. Our children had been fascinated with Lars; now they began to question his last name. Schmidt sounded Ger-

The F.B.I. and I

man, not Swedish. Maybe, they suggested hopefully, he was a German spy. I smiled my disbelief but the game of cops and robbers had given way to "German spy" and the children took the greatest pleasure in reporting to their companions that Mommy had a friend who undoubtedly was a spy and when he was caught we'd all have a pretty exciting time of it.

The day after America declared war on Japan, I started a war job. I worked for the Army Air Force in a supposedly secret location on West 18th Street and although we conscientiously got out of our taxis a block or two away, every taxi driver soon came to recognize the building. This was headquarters for the Interceptor Service. The flight of every plane was recorded on huge area maps and the volunteer workers known as plotters were in constant telephone communication with the outlying areas, where observers or spotters scanned the sky and telephoned immediate reports on every plane they observed. There was another and even more secret section where the reports of planes approaching the United States from any direction were picked up by radar. This was considered classified information. I worked directly with the army in the section known as Intelligence and Public Relations. At first I was there two or three days a week; bit by bit it crept up to full time. I was jack of all trades.

I edited a small house-organ magazine intended to disseminate information, attract new volunteers and keep up the morale of volunteer workers and army personnel alike.

I was a ghost writer for anyone with the rank of major and up; and took a snobbish delight when the speech was written for a general. Mostly, the officers were easy to work with and they were so delighted to be relieved of the duty of writing their own speeches that they would have accepted any first draft. One of them, however, who couldn't put three words together himself, suggested that I use more flowery language

Occupation: Angel

and plenty of quotations.

"Don't forget," he said, "I've had a classical education!"

Another officer had a real talent for ruining a speech by adding a little something of his own. One evening he was scheduled to make a speech in Tuxedo, where we were desperately short of volunteers to man the observation posts. I had worked very hard on the address which was aimed to persuade the solid, middle-class group of women who lived in the surrounding area to volunteer their services. The officer read my speech all right but he worked out an extra introduction of his own which he boomed out: "Ladies of Tuxedo Park . . ." (then in a lower, sloughing-off pitch) ". . . and women of the village."

"Your speech didn't pull," he said afterward, "they just sat there hard as stone!"

A short time later the same officer was scheduled to speak to the noontime crowd of women shoppers from the corner window of a Fifth Avenue store.

I urged him to try the same Tuxedo speech, with a few changes for more local appeal.

"And please, sir," I urged him, "stick to the opening salutation, 'Women of New York.'"

He promised, and he kept his word. As a speaker he was not ineffective, although he employed a style to which I was not partial. He was prone to use vast, proconsular gestures and he spoke very slowly and deliberately, which gave an air of consequence even to statements which had no importance.

I watched from the crowded sidewalk, wedged between a lamppost and the rather effete young man who had decorated the store window especially for the occasion. We each had our separate interest and pride in the event. I observed the faces around me and listened for comments. I was content with the women's air of absorption; the speech had gone well and should have been drawing to a close but the officer was apparently

The F.B.I. and I

pleased with the reception too, and he was loath to let go. The urge to hold his audience a little longer was hard upon him. He took a deep breath and extemporized:

"And in conclusion, may I say that I have no doubt about how you will respond to our call for volunteers because when duty calls, the women of New York always have answered. In fact, they have done more . . ." (clearly he was graveled for a superlative, but not for long and he finished with a triumphant rolling of his *r*'s) ". . . more than p-r-a-c-t-i-c-a-l-l-y any other sex."

For a split second there was a tremendous silence; one that it seemed would be impossible to repair.

"Oh, dear God, he's done it again," I wailed silently.

The window dresser next to me was the first to recover. "At last, *we* are getting recognition," he squeaked, making a gesture of perking up his coat lapels.

A wave of snickering swept the crowd, all the stronger because of the attempt to suppress it.

"Well, I've tried that speech twice now, and I tell you it just isn't any good," the officer complained to me later. "It gets away from me."

One of my jobs was to serve as liaison officer between newspaper reporters and the Interceptor Service. The newspapers had to be fed appealing stories that would entice women to volunteer their services. We had to disclose enough to the feature writers to excite their interest; leave enough unsaid to make them feel that everything was top secret and important. When I would take a writer on a partial tour of the building, I found that the more floors I said were not open for communication to anyone who had not individually been cleared by the F.B.I., the bigger spread we were likely to get. I did not bother to mention that aside from my initial, cursory clearance, I had been given full access to all departments and had been working with classified material for many months before any-

Occupation: Angel

one remembered to clear *me*.

I was supposed to know all aspects of the work and so two days a week I had to be downtown at 5:45 a.m. to check the six o'clock shift of the volunteers. That meant that on cold winter mornings I had to get up at four-thirty, that loneliest and eeriest hour when night was gone but daylight was reluctant to approach. There was a sort of toneless residue of black, more fearful than the dark itself.

It was an hour when one needed little comforts. I always meant to allow myself enough time to make coffee; I never did. I always meant to take the subway but at the last minute I'd eye the ghost-colored, deserted streets and decide to take a cab. That way I could stop at one of the all-night coffee spots downtown and have a quick cup of hot coffee at the counter.

It was a tough neighborhood and one morning at five, an overfriendly drunk joined me at the counter. I was more embarrassed than frightened but I felt easier when my taxi driver, who had observed the problem, slid into the seat next to me and said: "Sorry I'm late."

The drunk assumed that the driver was my escort and ambled off.

"You shouldn't come into these dumps alone," the taxi driver advised me. "The next time, just ask your driver to have coffee with you. It's safer that way."

A few days later, I remembered his advice and I said to my driver—or to that anonymous back which is all one usually sees: "I'd like to stop at the lunch counter on the next block; and, if you don't mind, please come in and join me for some coffee."

He slowed down his cab and turned three-quarters of the way round and looked at me. I in turn got a good look at him. He had a cauliflower ear, a squashed nose and an ugly scar for good measure. He was a veritable caricature of a thug. I wondered uneasily if that other taxi driver's advice was really as sound

The F.B.I. and I

as it had appeared at the time.

"Lady," he said in a highly disapproving tone, "I don't know what you have in mind, but I'm a respectable married man!"

After that, I carried a thermos with coffee from the night before.

For the first time in my life I understood the position of the tired businessman. Howard and I went out much less often; I no longer felt it necessary to see every show that opened, which was just as well. I was sleepy most of the time and I could not have stayed awake at most of them.

Wartime is not a time to inspire good, creative writing but even in a period when a certain degree of mediocrity would be understandable, the low ebb of the season 1941–42 was remarkable. For the second time since their formation, the members of the New York Drama Critics' Circle and the Pulitzer Prize Committee refused to name a best American play for the season.

There was little I could recommend to my Swedish theatrical partner but, with ten more productions than the season before, he found my paucity of recommendations difficult to understand. Time and again he would cable inquiring about a play.

Clifford Odets, who had created considerable interest in Sweden, had come up with a new drama called *Clash By Night;* Billy Rose produced it; Tullulah Bankhead, Joseph Schildkraut and Lee J. Cobb played in it. Surely, Lars pleaded in a cable, *that* must be a possibility. But the play had been a disaster and I cabled back, curtly advising him to avoid it and also to skip a quick failure called *Brooklyn, U.S.A.*, plus several other plays he had inquired about.

However, shortly afterward, I did secure the Swedish rights to *Arsenic and Old Lace* and to an old play of Elmer Rice's called *Black Sheep*.

The first year I still had a backlog of plays I had read and

Occupation: Angel

on which I felt qualified to make decisions, but by the second year of the war my life had changed. I was reading and seeing fewer plays and it was becoming increasingly difficult for me to find time to work into my busy schedule the tedious, mechanical work of contacting producers, getting options, endlessly awaiting confirmation from Sweden and cleaning up the other details. Some of the letters we were writing to each other were lost at sea, others arrived weeks and even months later. There seemed to be no continuity—no hope. I could not go on, and so with regret I terminated the partnership but we maintained the friendship. After that our letters were sporadic and were more like essays than letters: cries of protest not only at the horrors, but the painful dreariness and futility of war.

One evening a stranger came to the door and asked to see me alone on a very important matter. He identified himself as an investigator from the F.B.I. and supplemented his claim with a full list of credentials, including a photograph of himself. Considering his appearance, I thought it would have been kinder if the Bureau of Investigation had let him skip the picture. He was on the tallish side without really being tall; thin in a way that made one suddenly recall the word "meager." His large ears curved as if to catch every sound; and his pale blue eyes stared as if they did double duty and were used not only for seeing but also to supplement that all-important hearing. He had as little chin as it was possible to have, and he held his thin lips in a pinched fashion as if he were in pain but being very sporting about it.

We went into the library and closed the door. I assumed that the call had something to do with my war job and I felt a little thrill of pride; they must trust me enormously, I reasoned, for the F.B.I. to turn to me for help.

I inched my chair over a little closer to the investigator and kept my voice pitched low as I asked him what I could do for him.

I could answer some very important personal questions, he

The F.B.I. and I

said. He asked me if I recalled the usual, routine questions I had sworn to when I secured special clearance for my work. Then he wanted to know if I *still* maintained that, to the best of my knowledge, no member of my immediate family and no relative was residing in a belligerent country.

As an orphan with singularly few relatives, I did not have much trouble keeping track of the ones I knew. I reaffirmed the accuracy of my oath. Suddenly I was aware that my cozy, conspiratorial tone was not appropriate to the conversation. The investigator and I were not partners in anything, but were quite clearly opponents.

In an icy, unfriendly tone he told me that it was only fair to warn me that the slightest deviation from the truth would be used against me.

"Now, Mrs. Cullman," he continued, "through your work with the Air Force, you are acquainted with our system of radar control, are you not?"

I nodded.

"And you are intimately and precisely aware of the location of our air fields, and the number of interceptor fighter planes located in each area?"

Again I nodded. The next question was a bombshell.

"Were you using this knowledge when you warned someone in a foreign country against approaching or possibly invading Brooklyn?"

"This is absolutely ridiculous!" I protested when I could find my voice again. "I have no idea what you are talking about."

The conversation had taken on the chilly quality of a nightmare and the fact that I didn't understand it did not seem to put a stop to it.

"Who is Lars Schmidt? . . . And what is your precise connection with him?" the F.B.I. man wanted to know.

For a moment, I could hear the children's lilting chant: "*Mr. Schmidt's a German spy . . . Mr. Schmidt's a German*

Occupation: Angel

spy. . . ."

"Mr. Schmidt's a . . . a . . . sort of former business associate," I began lamely and then, ashamed lest my hesitation imply doubt, added firmly, "and a close, trustworthy friend." There! I had taken my stand.

"What does he plan to do with the lamb and rice he secured through you?"

"*Mr. Schmidt's a German spy . . . Mr. Schmidt's a German spy. . . .*"

"Lamb and rice? I . . . I don't know. I mean, I had nothing to do with it; I don't know anything about it." Small beads of winter perspiration were forming on my upper lip and my spine felt cold. Could I possibly have been used as a dupe?

"You say he's 'a sort of former business associate'; we have a copy of a cable you sent to him confirming the consignment and yet you 'don't know anything about it.' Do you wish to change that statement?"

I shook my head, too numb to speak.

"Do you also deny having secured poison for him?"

"Certainly I deny it! I never shipped rice or lamb or poison or any other commodities to Sweden . . . or anyplace else. It wasn't that *kind* of a business. He's a theatrical producer and all I ever got for him were the rights to plays—and not so very many of those. Why would I send him poison?"

Then with a wild attempt at a connection, I said: "Poison! Could that possibly have been 'arsenic'?"

My inquisitor pointed out that arsenic was considered poison, and there was sharp antagonism in his voice; but I was overcome with the almost hysterical laughter of understanding and relief.

"It was arsenic—it was *Arsenic and Old Lace!* I *told* you I bought plays for Mr. Schmidt. And *I've* got copies of my cables, too."

I kept a file of all the Swedish business in a desk next to where I was sitting.

The F.B.I. and I

"HAVE SECURED ARSENIC," I read. (Me and my prudent abbreviations!) "AVOID BROOKLYN CLASH BY NIGHT" and "BOUGHT RICE'S SHEEP." I clutched the cables triumphantly. My investigator looked blank.

"This Rice isn't a grain," I tried to explain, "it's Elmer Rice, the man who wrote *Black Sheep*. And *Clash by Night* is the title of a play, so is *Brooklyn, U.S.A.*"

On I went, translating one cable after the other, but still there was no dawning light of recognition on the investigator's face.

"Do you mean to tell me you've never heard of any of these? Don't you ever go to the theater?" I asked.

"Never!" He said it proudly, as if this were infallible proof of his integrity. "And if this is some kind of a code, it'll be broken down very easily."

But as he stood up, preparing to leave, I could see that the strength had gone out of his first conviction, leaving him without an established guilt upon which to rest.

"I may see you again," he said as if loath to let me off.

As the front door closed, Howard came into the room. He was curious to know what the visitor had wanted.

"He wanted to know if I were a spy—in league with Lars Schmidt."

"Schmidt the German spy." Howard gave me a brief smile. "No, *seriously*, what did he want?"

As I told the story, Howard was convulsed with a mirth I wasn't yet ready to share.

"You'll find it funnier in the morning," he assured me. "Too bad the kids missed it!"

And then he added solicitously: "You really look shaken. Better take a drink."

I tried to pour myself a brandy; my hand was trembling. Howard took over and poured the drink for me with the warning: "Careful, that's my favorite decanter, Swedish glass, I think—and I wouldn't dare cable for another to replace it!"

13

Larceny and Hot Ice

Probably *Life With Father* first whetted our desire to own or partially own a theater. When Lindsay and and Crouse booked their show for the Empire Theater, they were asked if they would be interested in buying the house. It could have been had at that time for about $215,000.

Following the depression of the early thirties, the lights of more and more theaters went out—and stayed out. The dry dust of disuse haunted these empty houses, which were falling into the hands of reluctant bankers. However, the Empire still was in good condition and it boasted a great tradition. It made one think of the earlier, romantic days of the theater when quite a few people with less direct interests than producers or chain theater owners fancied the idea of owning a

Larceny and Hot Ice

playhouse. Ownership carried an air of prestige; it was an exciting venture; and, according to rumor, there was real gold to be dug out from under the seats. In those days, more than one actress felt that a theater made an acceptable and appropriate little gesture of friendship from a wealthy protector—a tribute to her professional and private talents.

When the Empire Theater was offered, Lindsay and Crouse talked it over with Howard and all agreed that it would be better to pay rent and not become involved in the mechanics of owning an expensive bit of real estate which might be doomed to lie fallow a good portion of each year. The trouble with this conservative and reasonable philosophy was that the show in question, *Life With Father*, lasted longer on Broadway than any other play—and spent the first six of the eight-year run snugly ensconced at the Empire Theater. Halfway through the second year, the rent had amounted to more than the original asking price for the house. This knowledge hatched in the back of all of our minds leaving us, if not eager, at least open-minded.

Then another theater ran into financial trouble and was reluctantly taken over on a foreclosure by a bank. Theater owners who were logical customers proved skittish about buying it. One group, however, offered to operate it for the bank along with a chain of theaters they already owned and operated. In a short time, the bankers were even more unhappy: expenses seemed extraordinarily high and the bookings were not the most remunerative. At any price, they wanted the theater off their hands.

So, Howard Lindsay and Russel Crouse formed a five-way partnership with Leland Hayward, Elliott Nugent and Howard; they bought the theater and, in a burst of gallantry, agreed to make their wives the stockholders.

The deal was consummated and the theater became our property with the speed of lightning. Too late, the operators protested that if they'd known the bank *insisted* upon a sale they would have bought it; we were the new and very proud owners.

It must have been that pride of ownership that sent us on an immediate, housewifely survey of the property; and downright curiosity that made us insist upon opening a locked storeroom. There we found no murdered predecessors but an equally surprising cache—a roomful of toilet paper! There was row after row, stack upon stack. It was impossible to estimate the exact amount but the supply was infinitely more than any possible calculation of actual use could have suggested. Everyone disclaimed knowledge of, or responsibility for, the loot. The mystery haunted me, especially at unlikely moments when my mind should have been elsewhere. I was so fascinated that I repeated the story—inappropriate though it might have been—at a luncheon. A highly knowledgeable producer was present and he smiled at my bewilderment.

"But now, *really*," I continued, "can you imagine anyone stocking *one* theater with a supply that might last fifteen or twenty *years?*"

"*Or*," our cynical friend suggested, "which might supply their fifteen or twenty *other* theaters for one year each?"

"Do you mean to say that anyone would buy a year's supply of paper for all of his own theaters and charge it against *one* theater he was *managing?*" I was aghast!

"No! I did not say that!" he corrected. "I wouldn't think of *saying* it—it's a libelous statement and I'm a very cautious fella. But if there's a better guess . . . *you* make it."

I never could!

* * *

Having established a matriarchal ownership of the theater, Howard Lindsay became chairman of the board, the one male in our otherwise feminine coterie. I had assumed that any business problems or reports would be handled on an easy, telephone-call basis and that we would all respond to Mr. Lindsay's call with a variation of: "Oh sure, Howard, whatever you think best." But I had not reckoned with an actor who by then had played a successful businessman for five straight years. Even when Lindsay formally called a meeting of the stockholders, I think we all thought of it as a playful gesture. We were quickly disabused of the idea. Lindsay looked embarrassed when we kissed him (although we were all on a cheek-kissing, first-name basis) and invited us, collectively as "Ladies," to sit down. I waited for the joke that would dispel the formality; it was not forthcoming. Suddenly, I realized that, although this was not a matinee day, Howard Lindsay was still playing the role of Mr. Clarence Day attempting to explain the business structure to a composite, less than bright version of Vinnie, the wife with a butterfly mind.

Maggie Sullavan (who was then Mrs. Hayward) must have caught on at exactly the same moment I had: our eyes met in surprise and amusement and then she leaned over and, in a hoarse stage whisper, inquired where I'd gotten my suit.

I whispered back the name of the designer.

"I thought so—one can always tell."

Lindsay a-hemmed loudly and we both made the embarrassed, settling-back gestures of children who have been caught not paying attention.

Presently I risked another look at Maggie. Her legs were crossed and she was swinging one foot rhythmically. I took the cue, pointed to her shoe and made a finger circle, indicating perfection.

She beamed her pleasure and mouthed conspicuously: "They're Ferragamo. Not from Italy." Then, in elaborate pantomime, she indicated, "They have a representative right *here*." From the rapid, jerky pointing of her forefinger, one might have thought he was right in her lap.

Occupation: Angel

All eyes were on us. Maggie had attained her target but she was too good a showman to overplay it. We sat back, a picture of scatterbrained innocence.

Howard Lindsay fumbled and repeated himself; he was shaken, but not yet brought down.

"And now," he said, "I think we should discuss debentures."

The opening was perfect. "Oh, good," Maggie said happily. "Oddly enough that's exactly what Marguerite and I were discussing, just a little while ago."

I nodded enthusiastically.

Doubt that any of us had even heard of the term "debenture" was implicit in Howard's bewildered expression.

Maggie plunged on. "We were saying that if the benches—well, actually we said the seats—we were saying that if they had to be reupholstered—and they do look shabby—why should they always be done in that dark, plushy stuff? We thought a bright chintz would be terribly gay." She looked to the other girls for approval. They seemed slightly confused.

"Chintz," Maggie continued judiciously, "doesn't always wear so well, but if we *quilted* it . . ." Her husky voice rose on a questioning note.

For the moment, everyone forgot Lindsay entirely and agreed that quilting it could make all the difference.

"Ladies," Lindsay roared, and I was surprised that he didn't say, Oh, God!, "Ladies, I said *debentures*."

My husband says we perpetrated the worst pun in history, and I consider him an authority on bad puns.

* * *

Part ownership of one theater whetted our appetites and despite some of the hazards of theater-owning which we were beginning to discover, when the Hammerstein Theater, which included a fourteen-story office building, was for sale, Howard was interested. The asking price was modest and the offices alone would carry the building, leaving the profits from the theater sheer velvet.

Howard immediately offered our four partners in the other theater a participation; only one accepted. And then because the name Hammerstein was such an obvious association (the house had been built by Oscar's grandfather), he offered a piece to our friends and business associates Richard Rodgers and Oscar Hammerstein. They refused, on the basis that it might be wiser for them to refrain from mixing the business aspects of the theater with the artistic and creative end. This should have given us a clue because, when it comes to business dealings, Rodgers, a man of many talents, is about as vague and artistic as a precision missile. He was more experienced than we were and, like a wise parent, he let us find our own way.

When we bought the theater, it was being used as a television studio by William Paley of Columbia Broadcasting Company; Paley in turn then bought the Alvin Theater but, by the time he got an estimate on restoring the Hammerstein (which he was obliged to do), and moving all of his technical and expensive equipment to the new location, he hesitated and a compromise was reached. We kept the offices and rented the actual theater part to Paley again—and for the same sum he rented the Alvin Theater to us. We paid each other $100,000 a year rent as, apparently, a simple swap would have been unbusinesslike.

The two theaters gave us an added consciousness of having gone beyond the bounds of angeling. Each new play that I read now represented more than a possible investment—it also was a possible tenant for one of the theaters, and as such had an even greater potential (or so we thought) for profit.

There was one thing that bothered us. Bit by bit we were catching on to the fact that our original group of partners was not typical of the business. In joining the coterie of other the-

Occupation: Angel

ater owners, we were not exactly joining a group where everyone took honor and probity as a matter of course and abided by the Marquis of Queensberry's rules.

One Broadway theater owner, in fact, is as well known for the stories dealing with his outlawry as Goldwyn is for his malapropisms. These stories are never told with the slightest hint of censure; on the contrary, they are related with affectionate indulgence.

Some years ago a producer with whom we often invested had made a deal with this character. The arrangement called for a rent of 30 per cent of the first $20,000 gross intake, 25 per cent of anything between $20,000 and $30,000, and in the event that the show ran *above* that, the theater's rental then would be a straight 25 per cent of the whole. Only an immediate, smash hit with every ticket sold for every performance (an unlikely circumstance) would bring so high a gross. But to the surprise and pleasure of the producer and angels, it quickly became evident that the show would hit the $31,000 figure, which meant a reduced percentage for the theater owner, more money for us.

The young producer was in a fine mood that first Saturday morning when he dropped by at the theater to gloat over the advance at the box office. Then he heard a great banging inside the theater and upon investigation found two workmen busily removing the two far seats on more than a dozen rows.

"Boss's orders," they replied to his frantic urging to stop.

The theater owner confirmed the orders and blandly stated that he'd been worrying for some time that the sights on those seats were not good enough; it was unfair to the customers.

The producer made some rapid calculations and realized that just enough seats had been removed to keep him with certainty below thirty-one thousand! So the producer, who is a resourceful fellow, decided that although it then was customary to charge $4.80 week nights and $6.00 over the weekend,

Larceny and Hot Ice

the popularity of the show might warrant an increase to $6.00 *every* night. His coup was successful and it was apparent that he could count on from $32,000 to $34,000 a week gross; and a rental of only 25 per cent.

I listened to the producer's story as if he had been describing the *Perils of Pauline* instead of the box office receipts.

"Well, in the end," I observed, "the old fox cheated himself out of one quarter of the revenue on those thirty seats he took out, because later he couldn't very well backtrack."

"Oh, sure he could!" the producer said. "Within a few days, he decided that the sights weren't nearly so poor as he had feared—and back went the thirty seats!"

One might assume that with less than thirty theaters left out of sixty-four, these playhouses would be a rich source of revenue. But such are the peculiar economics—or the hidden leaks—of Broadway that this is not always the case. We had to try it ourselves to find out.

Having the ability or the luck to judge plays was an asset, but there was more—much more. Suddenly I, who had so determinedly avoided all business details, became fascinated with the pure mechanics of running a playhouse. I had thought owning a theater would be like owning any other property which you occupied yourself or rented outright to someone else. But nothing is ever so clean-cut or definite in the theater; the very nature of the business precludes such a possibility, and the peculiar hocus-pocus with which it is invested adds to the confusion.

To begin with, a theater owner cannot rent his property to a producer on a yearly or even a half-yearly rental basis because no one knows how long the new play will last. And so the owner gambles on holding his playhouse for the new show and demands a two weeks' guarantee in advance and, for the run of the play, a rental based upon a varying percentage of the ticket sales (or gross intake). The contract will include a stop clause allowing the contract to terminate when the gross drops below an agreed control figure, at which point the owner is entitled to give his tenant a notice requiring that he move or close within two weeks.

In addition to deciding which will be successful, long-running plays and booking them in his playhouse, the owner must worry about a great many other things such as conforming to the building code, obeying the fire laws, and adhering to censorship regulations which often are given the most surprising interpretations.

For some obscure reason, it is the theater owner and not the producer who bears the major responsibility if the play contains obscene language, or if it infringes upon a moral code which is just as unspecific as most rules of the theater. And if the Commissioner of Licenses decides that the public's morals are being endangered, not only may the play summarily be closed, but there is the added threat that the theater itself may be padlocked for the period of one year.

The theater owner is responsible for supplying ushers, attendants, a house electrician, carpenter and stagehand (although these three jobs, at the insistence of the unions, eventually will be duplicated by the production's electrician, carpenter and stagehands), a house manager and box office personnel, known as treasurers.

Thus, the producer of a play finds himself in the unique position of having rented space in which to show his wares without permission to consummate the sales, because the actual handling of the ticket sales is in another's control. This, above all else, serves to fuzz what should be a clear-cut line of responsibility and, in case of infractions, such as an investigation into the ticket-scalping racket, leaves it open for the theater owner to blame the producer and vice versa.

This question of ticket scalping and who gets the "hot ice" periodically comes up, raises a flurry and dies. Then, phoenix-

Larceny and Hot Ice

like, it rises from the ashes and repeats its exact and collusive pattern of behavior.

During an official and valiantly publicized investigation about fifteen years ago, thirty ticket brokers either turned in their licenses or had them revoked for a technical violation of bookkeeping. For several days it was a bit difficult to get hot tickets, then the boys were back taking care of their old customers. Sometimes they used a new name, perhaps an abbreviation of the old one, and although the telephone number might have been changed, the voice on the other end of the wire was strangely familiar; by coincidence, the delivery boy was identical. There was one difference: the price of the tickets was slightly higher.

Perhaps because the theater is not a basic necessity, or perhaps because theater tickets so often are considered a deductible business expense, few people are much concerned, and they seldom realize how much money the scalping racket involves. The difference between the box office take on a hit show and what was actually paid on the outside for the tickets may run from $5,000 on a straight show up to $20,000 a night on a hot musical. This could conceivably add up to over $2,000,000 a year—on *one* show!

No wonder investigations die early. There is money enough to kill them.

In the theater, a land where pilfering seems to be transmuted into a legendary, puckish sort of joke, and truth is stretched beyond the possibility of recall, it is difficult to apply the usual standards of honesty and integrity. But the dishonesty seemed so blatant to me and the system so obviously bad for the future of the American theater that I was filled with a high, crusading spirit, determined to bring justice where no one wanted it.

The first step of my plan was to get facts and evidence and write an article for the Sunday magazine section of a newspaper with national coverage. Once thrust into the limelight,

Occupation: Angel

I felt that the evidence could not be covered up or ignored. I had an immediate ally in the editor, a man of infinite courage and integrity, for whom I had written before.

Considering the fact that I had no official authority, I still managed to get quite a bit of information. I talked to people who were unctuous, smooth and told nothing; and I talked to those who, through indignation, were communicative and revealing, but I had to promise them anonymity.

As a kind of test laboratory, I used a theater housing a musical hit. Of the nine hundred orchestra seats, one hundred in prime locations at each performance were termed "house seats" and were reserved for people associated with the show and the theater ownership. Four hundred were allocated to ticket brokers; the remaining four hundred were *supposed* to remain at the box office to be sold to the general public. However, it is a commonly conceded fact that the box office is the *one* place where they do not sell you tickets—or reasonably *current* tickets—for a hit musical show.

And so, if the tickets are not sold at the box office, we must accept the obvious inference that brokers and scalpers secure more than their supposed allotment of tickets. But from whom? Anyone who has access to tickets is a possible suspect but he must have access to the *best* tickets to get into the real money.

A certain percentage of tickets may find their way to the scalpers through the treasurer; and whatever token of esteem and appreciation the scalper gives in return is known, curiously enough, as "ice."

I discussed this aspect of the problem with a charming and much respected man of the theater.

"Ah, yes," he sympathized. "It's not easy to find an efficient, dependable and reasonably honest treasurer."

"Reasonably honest?" I asked. "Does one qualify honesty?"

"But of course," he answered. "Now I have a very good

man, but I was careful in picking him." A note of modest virtue and acumen crept into his voice. "Several smart fellows applied. 'If this show is a hit,' I asked each one, 'how much ice do you expect to pull out a week?' The first two said 'None,' so I knew they were lying. The third fellow considered my question carefully, scratched a few figures on a pad, and said: 'Seventy-five to a hundred dollars.' I knew he was an honest fellow. He'd take, but he wouldn't squeeze, so I hired him."

* * *

I brooded over the eventual fate of the one hundred "house seats" which were withheld for each performance. The custom originated so that the producer would not be embarrassed by the unexpected arrival of someone who needed to be accommodated with tickets. But theater owners also must hold out a few seats for themselves, for their friends, and, as we discovered by experience, to spread good will among civic officials responsible for the exact location of a fire escape, the swing of exit doors, the precise moral or immoral connotation of a scene, or the number of people standing in the back of the theater at a hit show. We found that it sometimes took several pairs of good seats before some official could make up his mind about a possible infringement. But, on an average, we never kept out more than one pair of seats a night—sometimes not that. "House seats" are not to be confused with seats that are "on the house;" they must be paid for but at box office rather than scalper prices.

In the case of our laboratory theater, the withdrawal of one hundred seats in the best locations established a new high for needed accommodations. Half of the tickets went to the producers, stars, cast and professional associates of the show; the theater owner kept *fifty* for himself and his friends. At eight performances a week, that added up to four hundred seats a

Occupation: Angel

week or over twenty thousand seats a year! This would allow for a circle of friends so extravagant as to give pause to the most notorious playboy of the western world.

One could not help wondering if the theater owner might not have counted a scalper or two among his friends. In any case, the brokers' biggest supply of tickets obviously came from the four hundred set aside for just such a purpose. These tickets were highly desirable and it must have been quite a problem to decide who got them. If some of the brokers were favored in quantity above others and felt like making a weekly or monthly gesture of friendship to someone—well, friendship is one of the lovely things about Broadway.

Occasionally the illicit gains would go to someone's head and he would demand more than the brokers (who were themselves breaking the law) considered "legitimate." One scalper was filled with loud and impotent wrath; he had finished telling me about the theater on which I was conducting my research and in which he considered the payoff stiff but more or less standard. His complaint was lodged elsewhere:

"All right," he fumed, "so the dirty so-and-so has a hit in his theater and he knows we gotta lay hands on the tickets. 'Okay, boys,' he says, 'you want 'em, you can have 'em, but it'll cost you face value *plus* five bucks a ticket under the table!'

"Now that's downright dishonest," he observed with a fine touch of casuistry. "Why, that would bring the cost up to over eleven smackers a ticket; then, what with a little here and there for protection, the usual grease, office expenses and phone bills, I don't see how we can make a decent living. All us brokers ought to get together and refuse to pay—maybe dump all the tickets back in their laps!"

This momentary flash of insurrection seemed to exhaust him.

"Suppose some of the others start jacking it that high, that kind of money's hard enough to get back on a doll show but for Shaw or Shakespeare it'd be murder! A lotta my reg'lar customers'd sure be sore as hell. And what could I say to them: 'Culture comes high?' Geez! I don't mind coughing up a *little*, but *five bucks!* It almost makes me wish I was on the legit, then maybe I could file a complaint with someone," he finished wistfully.

My finished article met with so many editorial, blue-penciled question marks of fear that it had to be submitted to the legal department for clearance before it could be published. The lawyer who read it was personally enthusiastic about the article's information but stated professionally that it was libelous. Truth, he explained to me, was not sufficient protection against a suit. I could not blame the newspaper. Why should they risk a lawsuit when it was unlikely that they would be thanked or aided by government officials or even the income tax department? By mutual consent, the article was run in so highly emasculated a form that it was completely innocuous. No one was sued. On the other hand, in fifteen years nothing has changed.

14

Pound Wise

When we take our vacations in Europe, we either begin or end our trip with ten days or two weeks of theatergoing in London.

I enjoy this stay for the sheer personal pleasure I derive from fine performances, but when it comes to selecting plays we are willing to finance for America, it is often more difficult to judge from the live performance than it would be from reading the script. And the fact that the English and Americans speak approximately the same language is almost more of a hazard than if we knew from the beginning that the play would have to undergo a full translation.

I have to remember that a play which may boast four superb stars in London cannot possibly be cast that way in New York;

Pound Wise

financially it would be prohibitive, and few American stars are willing to sublimate themselves for the over-all good of a production.

I have to close my ears to the laughter and applause of the audience around me and remember that this is a British, not an American, reaction. I have learned that many plays whose very essence is English transplant or Americanize poorly. This is especially true of plays with contemporary settings and themes. Somehow, the further we go back in time, or the more abstract our approach to a subject, the closer we become.

A Man For All Seasons seemed to confirm this theory. Of course, it had a lot of other things in its favor. Everything about the play was brilliant: casting, setting, and the writing in which the playwright was wise enough to use More's own wit for much of the dialogue. We saw the play in London just after it had opened and several English friends in our party commented that, although it was great theater, they didn't suppose it would be anything for New York, which was more or less what I had just said about *Ross*. But A *Man* . . . seemed to have an extra dimension. I reasoned that a country has its own *little* heroes and its *great* heroes—the latter, like Thomas More, belong to the world.

When the play was scheduled for Broadway, we took as large a share as we could get. On opening night, as a kind of litmus paper, we took along our youngest son, Brian. I was torn between keeping my eyes fastened on the star, Paul Scofield, and watching the rapt, completely absorbed expression of our ten-year-old.

"Extraordinary!" I whispered to Howard. "For all seasons, for all countries, and obviously for all ages. We won't need to wait up for the notices on *this* one—we're in."

A Man For All Seasons was not only the most memorable British import of the 1961–62 season but was the single successful one. Seven others tried and failed; three of these, *Ross*,

Occupation: Angel

A *Passage to India* and *The Complaisant Lover* had enjoyed long and very successful runs in London. Each year, with one or more straight plays we prove all over again to ourselves how widely we differ from the English in taste. The English, in turn, have quickly closed some of our longest-running hits, such as *Life With Father*, *The Voice of the Turtle* and *Mister Roberts*. But apparently they scarcely can err in importing our musicals, which are superb by any standards. Americans in London are amazed to see that two-thirds of all the musicals playing there are imports from New York, but the remaining native third offer few prospects for a successful trip in the opposite direction.

One of the first things that strikes an American in London is the pleasure of theatergoing; next one becomes conscious of the ease, and then of the general interest. The theater is an integral part of living and belongs to the person of modest circumstances as much, and perhaps more, than it does to the wealthy. The comfort, convenience and taste of the patron come first. This, of course, is in direct contrast to our New York attitude, which apparently is based upon the theory that the customer is always wrong and might just as well be told off right at the beginning.

London has made the theater warm, personal, infinitely varied and easily available. Short of a sixteenth-century love potion, no one seems to have struck a more effective formula.

One year, the first thing I did in London was to drop over to the box office to pick up tickets which a friend had reserved for me. As the play was the newest and most popular hit, I felt a wave of pity for the man ahead of me who was asking if he could buy two good tickets—they must be down front because he was hard of hearing. Having brushed up against a few of the fellows in New York box offices, I winced. He's going to get it, but good! To my surprise, the ticket seller acted as if this particular sale were his chief concern in life. The two men had just settled on a date when the ticket seller posed a possible problem: "That's the evening following a matinee performance and so the play will start an hour later that night, eight-thirty instead of the usual seven-thirty. Are you sure that will be convenient, sir?"

The time proved convenient, the man bought his two tickets, paid fifteen shillings each—or $4.20 for the pair—and left with the air of a gentleman and an appreciated patron of the arts.

London is one place where tickets actually are available at the box office; and theater-ticket profiteering is rare.

Theater hour in London is another surprise to Americans.

"Why seven or seven-thirty?" they ask. And when Americans are told that it is because this is the most convenient hour, for the majority of the people, they look highly skeptical. Our own theater hour is not exactly inviolate; opening nights have been pushed ahead a half hour but that is for the convenience of the critics—not the public. Now eight-thirty just may be the most convenient hour for the majority of people in New York. But if it is, the chances are that it is purely coincidental.

The London theater hour used to be even later than ours, but during the war the hour was pushed ahead so that the streets could be cleared before the nightly air raids began. Later, people became used to the time and for many reasons it turned out to be more practical.

In London, an evening at the theater is a festive affair, but it is not necessarily a late night. It is the start of the evening for some—the windup for others.

Many of the plays give a conventional, mid-week matinee at two-thirty but on Saturdays switch to a performance at five-thirty and another at eight-thirty. That means four separate and distinctly different hours for performances. This arrangement is certainly a nuisance for the box office, a headache for

Occupation: Angel

the ad man and no picnic for the actors. It doesn't pretend to be. It is the theater's gesture of appreciation to the all-important public.

Arriving at the theater, one encounters noticeably less traffic congestion. More than 50 per cent of the credit for this may be charged up to the coincidence of geography rather than to careful theatrical planning. Some of the New York congestion is due to our mass passion for arriving within five minutes of curtain time, while in London the normal allowance is from fifteen minutes to a half hour. This custom is related not so much to inherent virtue as to the fact that the Englishman can settle down comfortably in his seat and have a smoke while he is waiting for the curtain, or he can meet his friends in the bar and have a drink, a cup of coffee, or something to eat.

During intermission, one rainy night in London, three of us were enjoying a drink in the theater bar; an Englishman whom we had noticed sitting directly ahead of us struck up a conversation. We learned that he came in to London once a month and always spent the evening at a show. He mentioned the play he was looking forward to for his next trip.

"But that didn't get very good notices, did it?" one of us asked.

"Notices?"

"From the critics," we explained.

"Oh, the critics!" He seemed to brush them off with an ash on his lapel. "I so seldom can be bothered in September remembering which ones liked a play they saw in June. Besides," he added with an air of modest accomplishment, "I almost never agree with them!"

We mentioned how cozy it was to have a drink right in the theater between acts.

"It's nothing very grand," he said, eying the room with its red mahogany bar and wobbly chairs, as if it were an odd thing

Pound Wise

he wouldn't like to pass off on a friend. "No doubt your own theater bars are much finer."

We explained that we had none.

"I suppose in that case most of you just stay quietly in your seats and have a smoke?"

"No smoking allowed, either; it's supposed to be a fire hazard."

"Why don't you fireproof things," he asked, ". . . you know, seats, rugs and maybe even the draperies?"

"We do. We even fireproof the props on the stage."

"And *still* you can't smoke?"

We shook our heads. It was too involved to explain that we hadn't suffered a theater fire since the days of our grandfathers, and that our fire prevention was modern; it was just our laws that were antiquated.

And as for not caring what the critics say, we have no possibility of this. We are so critic-ridden that if two out of the big three turn thumbs down on a play, it will be closed before we can forget their opinions.

Our friend of the intermission was representative of the loyal English theatergoer, who will plan to see the latest opus of a favorite writer, the present vehicle of an especially beloved performer, or even a current revival of a pet play. It doesn't have to be a smash hit, just so long as it provides him with a pleasant evening. Perhaps this attitude is not surprising when one considers that such an evening for him is neither costly nor troublesome. Nor, considering the trouble and expense a New Yorker is put to, is it unreasonable that *he* should hold out for the compensation of a great hit . . . or nothing.

This infrequency of the theater-going habit in America—and it is a habit—is responsible for our hit-and-miss system: the mad scramble for three or four of the hit shows, the cool indifference to the rest. The fact that the theater survives at all in America probably is due to the vital, almost dispropor-

Occupation: Angel

tionate enthusiasm and devotion of a comparative handful of people.

There is no doubt that England's theater is established on a much wider, more solid base. The Englishman considers it a vital part of his national culture. If the art of the theater can flourish on private capital, fine—but he expects his government to stand ready and willing to give aid if necessary.

At a time when England was eyeing our Marshall Plan money thoughtfully, hoping that it would stretch far enough to buy needed food and fuel, she was also—like many countries of Europe—carefully keeping watch over her artistic and cultural assets. And so, quite logically and openly, some of our money went to music and art, the ballet and the theater. This was a very happy arrangement; the arts flourished, and traveling Americans applauding wildly at the opera in Rome or Milan, the theater and ballet in London, told each other ecstatically that you couldn't beat the Old World at that sort of thing.

Long after England ceased to accept further financial help from America, the British Arts Council continued to assist the English theater. The system, broken down into its simplest aspects, works as follows: The Council stands ready to finance any accredited producer up to 50 per cent of the cost of production of any play which comes under the heading of furthering cultural interests. This, of course, offers a pretty wide scope and covers a revival of the classics, the importation of an American production, a play by a famous writer or the initial offering of a completely unknown one. The producer must form a separate, nonprofit-making corporation; the usual expenses are incurred and salaries paid, but no one may enter into a profit-sharing contract. The profit, if any, goes to the Arts Council.

Like any other angel, the British Arts Council sometimes breaks even, sometimes loses and sometimes makes a winning but, unlike most other angels, it is not concerned with making

Pound Wise

money, only with bettering the theater. Actors, producers and the public all benefit from this arrangement and the general standing and reputation of the British theater continues to increase.

However, every time a similar pattern for subsidizing the arts is proposed to our government, it is greeted with the wild invective usually reserved for an election year. The idea is said to be an unreasonable and frivolous extravagance which an overtaxed people could ill afford, or contrary to the best interests of American free enterprise. It even has been declared that once the government stepped in, culture would be out and graft and corruption would be rampant.

All the time that our government officials were making emphatic gestures with the right hand to amplify these objections, they conscientiously were doling out money with the left hand to further the arts. Only, these arts were not *American* arts and, therefore, by some strange alchemy of geography were supposed to be immune to our own local dangers! Not only is our theater not subsidized but—like playing cards, cigarettes and liquor—it is heavily taxed!

For years, London and New York have vied for the title of theatrical center of the world. Currently, we appear to be losing the race and I find it small nourishment to be told that we are losing freely, without government interference.

At the moment there is a refreshingly free exchange of ideas, of plays and of actors and directors. Both countries culturally are enriched by it but, with the uneven division of aid, we are headed straight toward financial and cultural loss. It is only natural for actors and playwrights to gravitate toward a stimulating center of activity; it is also natural for producers to try to find the best material for the smallest cost. And, in that respect, London becomes more and more attractive. Scenery, props and costumes—all transportable items—are infinitely cheaper there; even when one includes the protective

Occupation: Angel

fees we require for American scenic and costume supervisors, plus the cost of shipping, rebuilding or repainting of sets, English imports still are a bargain compared to our prices. More than one producer is considering the advantages of producing American plays first in London. According to the same reasoning, a producer may take a chance on bringing over a successful English play, star and all, because what would cost him $40,000 to import might cost him $100,000 if it originated in New York.

At this rate, signposts to the future are clear if anyone cares to read them and to consider how many hundreds of thousands of dollars could be diverted from our economy.

Perhaps someone should rephrase that old adage about being penny-wise and pound-foolish.

15

Take the Case of Mister Roberts

We have invested in nearly three hundred productions. There are some I scarcely can remember, some I want to forget, and some that remain as fresh and green as the season they opened.

When I try to explain special circumstances, such as the possible failure that turned almost miraculously into success, the marked difference between London and New York taste in plays, the general vagaries of theater, or the frustrations of dealing with the unions, the play I cite most often seems to be *Mister Roberts*.

It opened in February of 1947, so it certainly was not one of the first successes we knew; it made a great deal of money but not nearly so much as some of the other plays. It was by no means profound; it established no precedent for originality

Occupation: Angel

in the theater and, although it was not even my favorite play of all time, the name continues to crop up. Someone asks me what it's like to catch a show out of town and I find myself saying: "Well, the night *Mister Roberts* opened in New Haven . . ."

Or someone wants to know in how many ways we have been involved in any one production and we recall that once we had a three-way involvement: we owned the theater in which the play opened and ran for years; we had a straight share in the backing; and we had put up "first" or "front" money which gave us a piece of the producer's share.

Mister Roberts was not the first play in which we had become involved in "front money," that complex and most speculative aspect of angeling, when one invests in an idea rather than a completed playscript.

Richard Rodgers first started us on this newer and even deeper involvement. One evening Dick telephoned and said that he had just finished reading a book of short stories called *Mama's Bank Account*. Would I read the book and see if I agreed with him that with proper adaptation it could make a wonderful play?

I read *Mama's Bank Account* and was equally enthusiastic; and I was excited because for us this meant adding a whole new dimension to investing. It is one thing to decide—or perhaps "guess" is the better word—if a play in written form will successfully make the transition to a live performance; it is infinitely more risky to decide if a novel or a group of stories can survive transformation into the entirely different art form of a play, with its restrictions of time and space, and then the almost equally tortuous change from still life to live production.

Dick Rodgers suggested that Howard and I join him and Oscar Hammerstein in buying the rights to a play adaptation of the short stories. The next step would be to find a good writer whose talents would be in accord with the story. If we

Take the Case of Mister Roberts

couldn't get a playwright we liked, we'd waste our option money; if we got one we liked but his adaptation wasn't good enough, we'd lose our option money, plus the advance we'd have to pay the writer. But if all went well, Dick and Oscar would co-produce the play (a first for them in *that* field) and we would have the advantage of a greater participation in the possible profits.

We were off!

Our first choice for the playwright was John Van Druten and he was as charmed with the material as we were. It was no small feat to weave the individual stories into a play, but there was the hand of a master even in the first draft.

The play, renamed *I Remember Mama*, was an immediate and enormous success and it was our baby; we had loved it, nurtured it and brought it to fruition. It was all eminently satisfying.

From then on anything I read, including the labels on soup cans, I automatically evaluated for its stage possibilities.

Still, it was considerably later that *Mister Roberts* came into my life. At the time, the material was in book form. I barely had finished reading it when I had a telephone call from George Abbott telling me he wanted to talk to me about a book *he* had just read. It was, of course, the same book, and arrangements were made in lightning-quick time that we would split the cost of the option on the book which so intrigued us both. The author, Thomas Heggen, wanted to take a crack at dramatizing his own book and would work with Max Shulman, who had had some dramatic experience. We had to do it that way or not at all; we chose to do it.

After what seemed an endless length of time, a draft was ready and George Abbott had first chance at reading. He sent the material on to me without comment. I snatched at the manuscript, impatient, excited and expecting a job as sensitive and masterly as the one produced by Van Druten. As I read, I

Occupation: Angel

felt sick at heart; the humor seemed vulgar and tasteless, nothing was left to the imagination. I thought it a disaster. I telephoned George Abbott and my "hello" must have contained all the disappointment I felt.

"You think it's as bad as I think it is," he said without preliminary.

"Worse." I, too, could practice economy, if only of speech.

We were trapped with a bad play and no rights to offer it elsewhere for adaptation. Heggen and Shulman liked what they had written and George had the option of producing this version or of dropping our option and losing the money we had put up. The latter seemed to us the lesser of the two evils.

I was still smarting, not so much from the financial loss, as from the frustrated feeling that the book had the making of a great play, when I had a call from Leland Hayward. The dramatization had not proven acceptable to anyone; the rights to dramatize the book were once more available; this time *he* held an option on the dramatic rights and did I want to read the book and possibly go in on it with him?

"I know it by heart," I assured him, "but I'm a stubborn woman and if there's a new deal, I'll take a second try."

Josh Logan was also intrigued with the book and wanted to work with Thomas Heggen on an entirely fresh adaptation, buying out Max Shulman—then he wanted to direct it also.

This time I was even more impatient than before and I kept muttering that they had had time enough to write *The Forsyte Saga*. Finally, in what everyone else considered record time, the script arrived. It was utterly different from the first one; only the names remained the same, and the characters lived through some of the same experiences. I laughed aloud in a quiet room and although I rarely cry at anything but animal stories, I came perilously close.

This time there was no question: we were fully committed.

Take the Case of Mister Roberts

The next day I happened to see Josh Logan at lunch and I tried to tell him something of my delight and enthusiasm for the play. He looked vaguely pleased but not surprised. It was as if I had echoed the praise of a multitude. He had absolute assurance and confidence—a rare quality in any creative person. Despite this confidence Josh never allows himself to be sloppy; the better he thinks a play is, the harder he works to bring it to that final, polished perfection it merits.

Leland Hayward was regarded as a sound producer and raising the balance of the money presented no problem, except the possible problem of who to allow in, who to exclude. Before any producer may solicit funds, he must file an offering circular with the SEC (Securities Exchange Commission) under the name of the current production. Only after this has been cleared by the SEC may he take on angels. Most reputable producers have a group of investors who have backed previous shows and, if they should wish to widen the circle, it is not difficult to find out the names of potential investors. Indeed, the law supplies a first-rate, detailed listing for anyone who is interested, because by law a producer is obliged to file with the county clerk's office a list of all his backers on a show: names, addresses and the amount subscribed. He is also obliged to publish the list in two periodicals. One notice invariably appears in the *New York Law Journal,* the second in any local newspaper, preferably (from the producer's point of view) one of limited circulation because the advertising rates are cheap.

Neophyte angels should take note of the fact that their investments are public knowledge and any tendency to exaggerate their participation can easily be checked. Just as infinitely more ancestors are said to have been on the *Mayflower* than the ship possibly could have held, so we found vast numbers of people who claimed not only to have had an investment in *Mister Roberts* but also a piece of every other success-

Occupation: Angel

ful show on Broadway. Sometimes it is a temptation to say that you don't recall having seen a person's name on the listing. One enterprising angel has devised an even more subtle form of torture. He nods in a friendly manner, telling the blowhard that he remembers having seen his name on the contract . . . just above his own. The culprit then wonders whether the real angel is bluffing also—or knows him to be a liar.

At this stage the producer and director are involved in the most vital part of their work. In three and a half weeks of rehearsal time a static manuscript must turn into a live and vital production; a mood must be created and sustained and the characters must become plausible, understandable people in action.

By the time rehearsals have begun, the scenery, furniture, props and costumes have been planned and put into operation, although they may not appear until a few days before the plays opens out of town.

The three and a half weeks of rehearsal belong to the director, and during that time he must not only bring out the best performance of each individual actor but somehow transmute all the performances into a rounded, integrated whole.

Some directors will take the entire company and merely have the actors read the play and discuss its values for days at a time until he feels that they all see it from the same point of view. Another director may plunge into action more rapidly, putting the actors into place and "blocking" the scenes.

In the second phase of rehearsal, the director usually leaves the stage to the actors and sits in the dark orchestra listening and watching them move about among the rickety wooden chairs, wobbly card tables and hard benches that represent the eventual furnishings. Stage illumination is likely to be provided by a single glaring work light. During the early stages of rehearsal, the full play is not given but individual scenes or "bits of business" are rehearsed over and over and over again.

Take the Case of Mister Roberts

It is only toward the end that the actors progress to a "slug-through" with constant interruptions, and then on to a "run-through" of the whole play with few, if any, interruptions.

Each director has his own method of handling a play and working with actors. During the various stages of progress, I used to slip into the theater and watch *Mister Roberts* come to life.

Josh Logan is fascinating to watch in rehearsal. He is a tall man, not fat but big. He has the ability to make his body completely amorphous, capable apparently of assuming any size or form. Sometimes he would not only *tell* an actor what he wanted but also *show* him. And after a particularly kinetic bit of direction, he would throw himself onto a chair, half on, half off, arms or legs akimbo as if, for the moment, having no further need for it, he had discarded his body completely.

Later on in rehearsals, when he was watching but not actively directing, Josh would sit entwined in and around a seat, motionless except for his elaborate facial contortions and emitting a steady susurrus—almost like the hum of a beehive—composed of curious whispers, verbal shorthand notes to himself, and soft, appreciative chuckles.

Usually the playwright and producer attend only the first few rehearsals, when there is a possibility of cast replacements, and do not show up again until final full rehearsals, when cautiously they may make comments and suggestions. In the case of *Mister Roberts*, Josh Logan was both director and co-author and I daresay he merely communed with himself and nodded his approval. Whatever his system, he made magic; so great was his spell that as I watched a rehearsal on the bare stage, I could see the finished play, with props and costumes and all.

The actors also get used to playing with imaginary sets, and at the first full dress rehearsal they may go to pieces.

The next step in a play's progress is the out-of-town opening. This is a moment of great travail for all concerned. *Mister*

Occupation: Angel

Roberts had been booked early to follow the New Haven run with the Alvin Theater on Broadway, but all too often at the time of out-of-town tryout the producer is still negotiating for a New York playhouse.

It might seem like slipshod business to plan on a Broadway opening without a firm commitment from a theater, but that is just one of the ulcer-making aspects of producing. Many of the most desirable theaters on Broadway are part of a chain and the owner of the chain may have an "understanding" with the producer. "You're in, Charlie," he has told the producer with a comforting pat on the back, "we're going to take care of you in *one* of our houses, just give me a little freedom in which to maneuver." The inference is that the owner could book the play in his third or fourth most desirable house right then and there, but he is hoping to make room for Charlie in his *best* house. Charlie can't afford to be too insistent.

The fact that the contract probably will not actually be signed until after the theater owner has seen the play out of town is sheer coincidence. But then, if he thinks the new show's chances are good, he undoubtedly will find that the occupant of his prime or second best theater is falling below the control figure at the box office. "You know that we want you with us, Joe, old boy," he assures the producer of last year's hit as he arranges to switch him to a smaller house, "but previous commitments, y'know." On the other hand, if the theater owner feels less confident of the new show, Charlie will find himself booked in one of the chain's less desirable theaters. "Small," the owner says warmly, "but a little gem."

This is the time, too, when the producer must face, and possibly fight, the somewhat autocratic rulings of Local 1 of the International Alliance of Theatrical Stage Employees. Regardless of how many men the producer may *think* it will take to mount and run the show backstage, the union steps in and issues yellow cards to *tell* him. If the union tries to stick him

Take the Case of Mister Roberts

with too many men, an experienced fellow may have the courage to object and demand negotiations. But the out-of-town tryout is a period of such stress under any circumstances that it is not surprising if the producer fails to take on one more battle—especially if the show looks doubtful. And if the play looks like a potential hit, overloading to the extent of two extra stagehands may seem unimportant. It is likely that the producer has other, more pressing problems. Perhaps the cozy relationship between him and the author has been strained. At the first suggestion of script changes, based on the audience reaction, the playwright may begin to feel less appreciated than he had at first felt himself to be. And although he and the producer both are busy taking the pulse of the audience, they may come up with different conclusions.

Tension may develop as well between the director and the actors, who are not only giving regular performances but are also continuing to rehearse three or four hours a day. And opening night out of town is, in a way, more important to the actors than the Broadway opening, because at this point anyone is still expendable. Once the play has come to Broadway, it may sink or swim but it does so with the cast intact. A long run or a short one, they're all in it together. It is their out-of-town performance that cries: "See how good I am, how perfectly I fit the part! You couldn't do without me!"

The opening night audience out of town is an interesting blend of local theater lovers and a varying influx from New York. Some of the people have a professional connection either with that show or the business in general, but there are plenty of plain cash customers who are lured far afield in an urge to see a hit-in-the-making and to be the first to talk about it.

Just how many of the first-nighters have any actual business with the show and how many are disinterested spectators is difficult to judge—especially since so many of the latter pretend to be the former and the former the latter. Many of the professionals, in a vain attempt to camouflage their mission, act as though they're only along for the ride. This is especially true of other producers, any one of whom may be present because he is fearful that this play parallels one that he is planning to produce later in the season; if it does, the sooner he knows this and can either make necessary changes or drop his option, the better. Or a producer may feel that the star of the show is the perfect lead for a play he is planning. If the out-of-town opening looks like a flop, he will hold up casting of his own show because the star of this one may soon be free. Then again he may be less the vulture: he may be negotiating with the choreographer, the scenic designer or the director and this will give him a chance to verify his opinion and possibly close a deal before success swamps the candidate with other offers.

The night that *Mister Roberts* opened in New Haven, we were part of the infiltration; whether we could have been called professional or nonprofessional would be hard to say. At any rate, we had a three-way participation in the play. I was as confident of success as I had ever been with a show, which means that my confidence was tempered. I never feel 100 per cent sure and I found myself eying the people in the hotel lobby with the quick, nervous but careful evaluation I'd give my opponents in a tennis match.

At the theater the very air was frosted with the excitement of a hit about to be born. And, because both of us were a bit nervous, we reacted typically in our exactly opposite fashions: Howard walked around greeting friends and prophesying the greatest success for the play; I sat absolutely still as if my seat were a mold into which I had been poured and was slowly hardening; my breathing was rapid but shallow. All in all, I was braced to suffer the least possible pain in case of failure.

I glanced at the rest of the audience. As reactors, I figured

Take the Case of Mister Roberts

that almost 20 per cent could be automatically eliminated; they either were too busy working to react at all, or were so biased that they would laugh at anything. It was the other 80 per cent—the good, solid public—that concerned me. As a rule this out-of-town audience is infinitely more courteous than a New York opening-night crowd. These people respond out of their own inner emotions without glancing furtively around for verification.

The lights dimmed and I felt the familiar double heartbeat that signals the curtain's rise.

If anyone on the other side of the footlights was nervous, no one would have suspected it. Everything about the performance was right. Each role, no matter how small, was played with the assurance of a star; and the star, Henry Fonda, seemed not to play but to live his role.

Technically, the New Haven tryout was a difficult proving ground; the laughs were so loud and so long that they overlapped and it was impossible to tell which bit of dialogue or stage business really was getting the reaction. But what a happy fault! And, perhaps because emotions tend to swing a full pendulum, the almost hysterical laughter was followed by open, unashamed tears for the death of Mister Roberts. Even I, who had heard the sad news delivered no less than twenty times, was touched anew.

After the show, the theater lobby was like a reception room and everyone congratulated anyone who had even the remotest connection with the production, right down to the local florist who had contributed the potted palm tree. And so, I was not too surprised when Howard's brother, Joe, our relatively silent but enthusiastic partner, patted me on the shoulder.

"This time, my girl," he said, "you've really hit it! Front money—regular backing interest—and booked to run in our own theater where it ought to last for *years!*"

"We're lucky," I said. "It represents a big risk."

Occupation: Angel

"Nonsense," Joe returned generously, "it's more than luck. And don't worry about money; I'll risk it on whatever you pick. But there's something I've been meaning to talk to you about. You know I never interfere, but as an old-time, tough businessman, I'd like to suggest that you follow this exact pattern for all future investing."

Bewilderment must have been etched on my face.

"Don't you understand," he explained patiently, "less money in flops—few as they are—and more money in hits!"

Joe's argument was irrefutable.

We had made our hotel reservations and were prepared to spend the night in New Haven. Others of the New York crowd had come similarly prepared but from times past I knew that there would be still others who would optimistically plan to return to town after the midnight conferences had broken up but who would trail down to the lobby at four in the morning, exhausted, and demand a room for the rest of the night. If the night clerk should be unable to accommodate these people, they are both surprised and indignant, "I only want a place to *sleep*," they explain patiently, as if this put a new complexion on the room shortage, "and just for *a couple of hours!*"

On an opening night, the hotel apparently borrows the Yale honor system and gilds it over with a nice sense of discretion. At any rate, with hushed or hurried consultations and conferences in all the different rooms, it would be impossible for any house detective to keep track of who is making use of whose room.

The hotel manager passed us with the frozen, tortured expression that comes over his face on big opening nights. I nudged Howard and asked: "Shall I make an honest man of you and thank him for my bedjacket?"

Some months earlier we had gone to another opening and had also spent the night. In my early-morning packing I had

Take the Case of Mister Roberts

mislaid my bedjacket. It was not until several days later that I realized my loss and, when I did, I telephoned Howard at his office and reported it.

"It's the pretty white lace one that you gave me for Christmas —the *Arsenic And Old Lace* one."

This seemingly unintelligible description was perfectly clear to Howard, who had hit upon the unique idea of giving me presents of lingerie which were named for plays we had backed. For a normally conservative man he had come up with quite a few imaginative bits of frivolity: a wondrously sheer black chiffon nightgown called *Let's Face It* had remained in the bottom drawer; but a turn-of-the-century black velvet hostess gown called *Life With Father* had seen constant service; so had my *I Married An Angel* slip that sported a couple of appliquéd satin cherubs. He even had gone to the trouble of having a dagger with one crimson drop of blood embroidered like a monogram on a travel robe—for *Macbeth!* The Christmas present from two years past, a white gown and bedjacket trimmed with real lace, was my favorite—and now I had left the jacket behind.

"Don't worry," Howard had soothed, "I'm sure they must have it at the hotel. Miss Burkes," I heard him call, "please drop a note to the hotel in New Haven, tell them what room I occupied last Thursday night and ask them if a white, lace-trimmed bedjacket was found."

An answering letter came back promptly and in its few words there was a world of cynicism and discretion. The note said that, yes, indeed such a garment had been found and would Mr. Cullman like to have it sent to his office or would he prefer to supply the name and address of the lady and have it returned directly to her?

We decided against further harassment of the manager but Howard wore the pleased look of a man upon whom the title of "rake" has been conferred without impairment of his morals.

Occupation: Angel

After we had seen *Mister Roberts* that night in New Haven, we all knew that another hit was on its way to Broadway. The knowledge was an intangible thing and not actually based on the audience's reaction because all of us connected with the theater are grimly aware that, in the final analysis, an out-of-town audience is not a Broadway audience. And this fine line of distinction may be responsible for making a producer believe he's coming in with a hit only to find it doesn't make the grade.

When a play fails, one of the questions asked most often is: "Why did they bring it in? Couldn't they tell it was a flop when they saw it out-of-town?"

This is a good question but it requires half a dozen answers. Everyone connected with the show wants to bring it in.

A playwright would rather have a hit than a flop but some playwrights have first come to the attention of the public and received encouraging notices on a play that was a financial failure.

An actor may be privately convinced that even if the play is panned he, individually, will get good notices.

The producer at this point has spent nine-tenths of his budget and he is aware that if he returns the balance to his investors they will find it small comfort to have lost 90 per cent of their investment without ever having had a real run for their money. The possibility of a movie sale is another inducement and the producer and his investors will share in 40 per cent of this bonanza only if the play has opened on Broadway and run for twenty-one consecutive performances.

Everyone connected with the show recalls the plays that made the grade in spite of adverse criticism out of town and, after all, they tell each other, employing a classical cliché: "All this needs is a little tightening up!"

And so the play is launched, to face the critics and the audience.

Take the Case of Mister Roberts

No such cloud of apprehension hung over *Mister Roberts*; it came in with the assurance of a great hit and it lived up to expectations. We enjoyed every one of the usual additional bonuses of a hit: a good, fat movie sale; a second income from road companies; and television, radio, local stock and foreign rights. We also had the rare good fortune of having a hit show playing steadily in our theater for three years! As a theater owner this was great, but as a backer this was a mixed blessing. For the first time in our theatrical lives, we were encountering union troubles.

Most of the complaints against Local 1 of the International Alliance of Theatrical Stage Employees are petty, yet our resentment was all the more keen because, unlike the Musicians' Union, Local 1 has no unemployment problem to justify its actions. In fact it has more jobs than members! Consequently, "moonlighting" (the romantic expression which means holding several jobs at the same time) is quite general. One job seldom keeps a man busy. The union keeps its members happy by maintaining a tight rein on the admission of new members and is vigilant in seeing that all work is highly restrictive and unlikely to exhaust or confine the worker. Thus, if a man's sole function for an evening is to push a light button, he is classified as an electrician and under no circumstances may he also move a chair, because that job is reserved for a prop man.

To pay for these unneeded or unrendered services, despite the soothing term of "feather bedding" by which they are known, is always irritating. We faced a triple dose of it during the run of *Mister Roberts*.

The American Federation of Music, better known as Local 802, requires theater owners to declare, at the beginning of the season, if they want to operate their house on a "contract" or a "penalty" basis. Contract houses are obliged to employ four musicians *whenever* they have a show but the musician's pay scale is kept down to $120 to $170 a week. The penalty house

Occupation: Angel

pays musicians only when the rules governing the show require them (i.e., with recorded music four standby musicians must be hired) but the musicians get from $160 a week for a drama to $213 for a musical comedy. And since the union demands that a musical employ at least twenty-six men in a house the size of the Alvin, if we had booked a musical comedy, or even a little revue, running a contract house might have meant a saving of more than $1,000 a week on the music bill.

And so, our house manager quite wisely had advised us to register as contractual. Then, in February the Alvin turned out to be the best available theater for *Mister Roberts*. As for the little obligation to Local 802, well, it was a slight irritant but, after all, the show looked like a hit; and in three months the theatrical season would end and we could sign up on a different basis for the following season. Meanwhile, Leland Hayward would be compelled to carry four standby musicians for *Mister Roberts*. A little extra salt in the wound such a needless expense incurs is the fact that every time musicians are hired, one of them must act as the contractor, which means that *he* gets 50 per cent above the union scale. In return for this emolument, he acts as hiring agent and liaison agent in relations between the musicians, the union, and the show. Never mind if the musicians happen not to play and have no need for contact, he's there to serve in his unnecessary role—and, of course, there to collect the extra fee. And even if only *one* musician had been required, he would have represented himself to himself and gotten his 50 per cent extra. There is a strong flavor of Gilbert and Sullivan about the situation, but one needs a highly determined sense of humor to laugh. However, if we were going to laugh at all, that was the time for it —not later. Because later we discovered that not only was Leland obliged to carry the four musicians for *that* season, but he was stuck with them for the full three-year run of the play!

Take the Case of Mister Roberts

Even the Ancient Mariner had but *one* albatross around his neck.

We, as the theater owners, were not allowed to exercise the annual option of shifting the theater's category because the union said the house had a "continuing attraction." All of us were furious, but absolutely powerless.

For a while the four musicians used to show up and sit in the basement playing pinochle but they proved such a nuisance, sending out for sandwiches or packs of cigarettes, that they were asked to stay away. They got their checks by mail and they qualified for their annual union-guaranteed vacation by doing nothing. On the show's first anniversary Leland planned a party for the cast and asked the musicians to play; they were quite offended at this interruption of their routine but, after considerable grumbling, agreed to do it just that once! Hard as it was to pay them for not playing, the week that they were paid *overtime* for not playing on a holiday was the hardest of all.

There is little that the most astute producer can do to combat this problem and most of the producers concede that actually Local 802 tries to be cooperative; if they held to the exact letter of the contract, charges could be even higher. Of the local's membership of twenty-nine thousand musicians, scarcely 1 per cent hold jobs in the legitimate theater. No one wants to trample down the underprivileged but the theater, struggling for its own very existence, can ill afford to subsidize anyone else.

As I say, if anyone asks me questions about the theater, I usually answer: "Well, in the case of *Mister Roberts* . . ."

16

Then and Now

I was sitting in our library at home; a few scripts lay on the table but I was at my desk. On one side of a beat-looking typewriter was a folder crammed with old clippings from *Variety* and the theater sections of the morning papers; and almost spilling off the desk on the other side was a long, detailed statement from the office listing the names of shows in which we had invested, the year the venture started, original investment, investment liquidation, and finally the cumulative profits or losses on each show to date. On top of the typewriter and in my lap were several small pads filled with scribbled numbers and percentages. It was the beginning of June, the end of the 1961–62 theater season, and I didn't like the way the figures were adding up. I nibbled the eraser on the

Then and Now

end of my pencil and made a face: the least they could do was make erasers chocolate-flavored for people with problems. And we had a problem.

According to my figures, financially the bloom was off the theatrical rose. The previous year had been bad; this was even worse. At the end of the theatrical season, fifty shows had opened on Broadway, only ten of them had or would come out in the black; another seven never had gotten beyond an out-of-town tryout. That meant that the odds had risen from the traditional four failures to one success to almost six to one. And failures came high. Every time another musical flopped, down went an average of from $400,000 to as much as $650,000. The very thought of carrying one-fifth of such financing and continuing to back ten or twelve plays a year gave me goose pimples on the back of my neck.

I tried to cheer myself up by looking at the figures on some of the season's hits we had backed, but there were no old-time, fifty-to-one payoffs there; not even half or a quarter of such scoring to brighten the picture. Most of them would return one-and-a-half for one up to three for one. A few showed promise of reaching a higher scoring; and several other hits on Broadway would do as well or better, but we had not had an opportunity to invest in those. That was another thing that rankled! At least in other forms of gambling such as horse racing or roulette, no one says: "Sorry, but we're not taking any outside betting on the favorite in *this* race." Nor does a croupier decide that number twenty-seven is coming up so often that this time he's going to reserve it exclusively for his own family and a few fellow croupiers; or break the rhythm by declaring that the next spin will be reserved exclusively for one patron.

I went back to my notes. There must be a simple way of expressing the problem—the angel's problem, to be specific, as apparently everyone in the theater today is screaming about

Occupation: Angel

his own particular worry. We used to be able to invest less money, secure a greater percentage of the play for it, and wind up with proportionately better returns. But I felt that I had to express it in figures.

I took a fresh approach, drew a line down the middle of the paper and on one side I wrote THEN, on the other NOW. Under THEN I tried the figure $20,000 as a fair price for a well-produced straight play with one set. Assuming that the play was an immediate hit, I listed a gross weekly take at the box office of $21,000 and against that I charged off $5,000 as the theater's approximate rental and $6,000 as the weekly running cost, leaving a net profit of $10,000—roughly 50 per cent of the gross. I did not pull these figures out of the air; they were taken straight from old weekly financial statements sent out by the producers. The addition of a big star to the cast would not have increased the weekly "nut" by more than a few hundred dollars, scarcely enough to change the rough percentages.

Under the NOW side I started to put down $100,000, crossed it out and substituted the more accurate sum of $120,000; six times the old cost of production. It wasn't hard to estimate the gross box office receipts; theater tickets have doubled in cost and a sellout in the same theater (indeed, there *are* no new ones) would bring in twice the amount, $42,000. By and large the theater rental remains proportionately the same and would run ten thousand, five hundred dollars; this might lead one to believe that the profits also would be doubled, but such is not the case. The weekly profit would not be $20,000, but more likely $8,000.

Who gets the other $12,000? Anyone and everyone who can get his hands on it! But this does not work from the bottom up so much as from the top down.

The average playwright is content with his 10 per cent minimum guarantee. Not so the upper echelon, who may demand the usual royalties plus as much as 12 or 15 per cent

Then and Now

of the show's profits. A member of that group may also demand star casting for his play. In that case the star (or stars) is likely to ask for a minimum guarantee of $2,500 against 10 per cent of the gross. And staggering as a salary of $4,000 or $5,000 a week may seem, it is nothing as compared to the demands of the top musical stars, who have gotten similar salary guarantees against one quarter of the entire gross. The director, too, has learned to take a good bite out of a show. He gets a flat fee, usually from $2,500 to $5,000 plus 2 or 3 per cent of the gross and sometimes an additional 5 per cent of the profits. One prominent director managed the usual terms plus 20 per cent of the profits.

Overdemanding authors, actors, directors and unions squeeze the producer until *he* figures a way to get even. Waiting until the show makes a profit and then splitting it with the backers is no longer enough for most producers. They have upped their weekly office expenses from $100 to $350 (a few charge $500) and most of them clip off 1 per cent of the gross as a sort of consolation prize in case the play never really makes the grade.

In the end, everyone is nipping away at the profits except the hapless angel; and it may be that he is sacrosanct only through lack of opportunity. It is a discouraging picture because the very people who should nurture and protect the theater generally seem more interested in cashing in on success. It was all right for them to starve for a cause when they had no money anyhow. But once at the top of the ladder, the cause turns out to be personal. Thus what starts out as a true devotion to the stage often becomes distorted through a simple syllogism: The theater is the most important. I *am* the theater. Therefore, *I* am the most important!

There seemed little point in trying to sprinkle pink dust over the picture, it was too sharply and clearly drawn. I'll worry about it tomorrow, I decided, and deliberately set about empty-

Occupation: Angel

ing my mind.

The front door opened and then slammed shut. Howard gave me a chaste salute on the cheek and sank into a deep chair. "Lousy day," he observed, "but it's clearing and it looks nice from here."

Opposite his chair was a low table with a huge bowl of pink and mauve tulips; beyond, through the open French doors, a dozen or so matching flowers growing in terrace boxes gave an air of abundance, as if the terrace had yielded up the indoor bouquet also. The effect was one of my cherished, deliberate deceptions. A bright sun had come out after an all-day rain; the red tile of the terrace still glistened wetly. A sturdy wisteria vine was heavy with freshly washed white blossoms; raindrops sparkled on a flowering crabapple tree, thick shrubs and climbing vines. Inside, the room was a cool beige and white.

"I like this room," Howard sighed contentedly. "And although it's different, somehow it has the same stamp, the same feeling as our first little apartment."

"The planting?" I suggested. "And, of course, a good deal of the furniture is the same; my desk is as cluttered as ever, and I still drop open books and magazines around as if I were going to use them as markers to retrace my steps."

Howard's mind was still on the first apartment. "It's funny," he said, "after over twenty years, I still remember everything about the old place. It all started there."

A soft silence dropped between us, the kind that sometimes occurs when two people are so at ease with each other that they feel no need to speak but go off on quiet little journeys of their own.

I stretched and my mind curled around the memory of our romance, which was all tied up with the apartment across the street from my office. The courtship had had an inauspicious beginning, to say the least.

Then and Now

I was new in my job as public relations director of Bonwit Teller and exceedingly busy when I had a telephone call from a girl I knew who was doing publicity for the Roxy Theater. Her request was simple and straight to the point—a publicity tieup. The Roxy had booked a British film with a child star whose wardrobe had been designed by Schiaparelli. I was interested and we made a date to meet for luncheon at a little restaurant halfway between her office and mine. The day of the luncheon found me frantically busy and if I had had the courage I daresay I would have broken the date. What's more, I was in a poor mood to wait ten minutes for my confrère. When we finally sat down, I noticed that we were at a table for three.

"My boss, the trustee of the Roxy, is joining us," she explained. "I thought it would simplify things."

"Wouldn't it have been better to wait until we know a little more . . . ?"

"No. He's leaving tomorrow on a trip and will be gone for a couple of weeks. It's better this way; I've been wanting you to meet him anyway. He's quite a remarkable fellow."

I thought I detected a matchmaking glint in her eye and I resented it. A rather dashing ex-beau had shown up in town that morning and I had been obliged to refuse his invitation to lunch on the grounds that I had a business date. Well, I'd see that we accomplished as much as possible during lunch.

Howard Cullman arrived and I judged him (quite accurately) to be about forty-two; and decided (quite inaccurately) that he was spoiled and used to being catered to. He urged Martinis on us and brought exactly the distracting, social element to the conversation that I wished to avoid. With a kind of bulldog determination I kept switching the conversation back to business. Howard wanted to relax and talk. I was in a hurry; I wanted pictures, details and a release to use the clothes in a promotional tieup. My host and I clearly were

Occupation: Angel

at cross purposes; and our playful snipes at each other began to grow somewhat barbed. He made a few deprecatory remarks about career girls. I snapped back that I found them more interesting than women who were kept, but that that was undoubtedly a matter of personal taste. Out of pure politeness toward the now unhappy third member of the party, we made an effort. Howard volunteered the information that he had just moved into the apartment hotel directly opposite Bonwit Teller but he didn't face the store, he had southern exposure. I too faced south and we agreed that it was the most pleasant location. At least, I thought, we would not be obliged to face each other across 56th Street even though Cox-and-Box-like we would most likely be there on different shifts.

The meal might have turned out to be just another boring, quickly forgotten incident, but our host had a certain bulldog tenacity and seemed unable to drop the argument. We were discussing the reorganization of Bonwit Teller and, with an elaborate mockery of being impressed, he asked me if I were the youngest executive in the firm. I thought a moment and said that probably I was. I countered by asking him if he were the eldest in his. For a man who had not quite gotten over the surprise of finding himself over forty, this must have had the effect of a mortal blow.

All three of us looked at our watches simultaneously, gulped our coffee and departed.

It must have been a month or six weeks later that I stayed at the office late one evening. My small, blonde secretary generally took a very dim view of working overtime but that evening she was in a good mood, chatty and confiding. After listening to the third case history of the various men who were taking her out, I tuned her out and barely was conscious of the soft, pleasant hum of her voice. Then I was aware of a change in tone—she had asked me a direct question.

"I said," and she stared fixedly out the window as she spoke,

Then and Now

"that man I was telling you about seems terribly interested."

"That's nice," I said absently. "Er . . . which one?"

"The one on the roof opposite. He's been there about ten or fifteen minutes, wandering all around but mostly staring over here. Well, now he's waving."

"Pull down the shade," I suggested. "You shouldn't encourage men on roofs; they're like truck drivers, they encourage easily."

She pulled down the blind but the spirit had gone out of her work. Then the telephone rang. My secretary answered and there followed an odd series of chortles, giggles, a "Yes, I did, but I was told to." More giggles and then she announced, "It's for you! I didn't get his name but he says he's the man on the roof opposite, and he asked for *you*—by name."

In her voice was the inference that I was a fun-spoiling, pious fraud who trafficked in roof-top flirtations on the sly.

"Hello," I said cautiously, prepared for almost anything.

A warm, deep voice greeted me, announced that he was Howard Cullman and did I remember him?

"Of course I remember you!" My voice was encouraging, solid with recognition. "If I sounded confused it's because my secretary," and I gave her an arch look, "said something about your being a man on a *roof!* Silly of . . . You *are* on the roof?" I had lost face; and the brief welcome note left my voice. "What do you do, Mr. Cullman, fix radio aerials in your spare time?"

He was determinedly cordial. "Uh, uh! I can barely change stations. I've just moved upstairs into the penthouse apartment and my terrace has a fine view of your office. I could even see you from here in my living room if you hadn't had the shade pulled down! Say, do you know anything about penthouse gardening?"

I admitted that I didn't.

"Well, you'd better start taking an interest because from

Occupation: Angel

nine to five you're going to be staring straight at *nothing* or at whatever I plant! Come to think of it, you'll get more out of it than I will. How about coming over for a drink? If you're athletic enough, you could practically jump it!"

"Sorry, but I'm still working—and I've an early dinner date."

"Does that mean we're still mad? I could get even and plant ragweed if you're allergic."

"No," I laughed, "I'm not still mad, and no, I'm not allergic. But hold off on the ragweed and I'll come over if you'll invite me some other time."

"Tomorrow?"

"Tomorrow!"

It was too late in the season to do anything practical about the planting, but we continued to discuss it over dry Martinis all fall. In June, right after we were married, I had a high picket fence put up.

"It'll make a better background for the planting," I explained, "and although not naturally jealous by nature, now that I'm on the magazine, no longer across the street, and I can't tell who's in my old office . . ."

"Don't worry about protecting your interests," he assured me, "I'll never let you go! I'd never have the strength to conduct another such campaign."

And indeed, it had been a campaign in which he had employed generalship, seamanship, statesmanship, strategy and humor. He overlooked nothing and forgot nothing. It had not occurred to me that I was a prize worth such passionate endeavor but he interpreted my bewilderment as reluctance and redoubled his efforts, tilting with windmills and slaying nonexistent dragons as he went.

Later Howard loved to tell the children, then the grandchildren, and practically anyone who would listen, stories of how he had won me. As the years passed, the stories became highly apocryphal. I was a princess in a tower whom he had

Then and Now

captured by accomplishing the most formidable tasks and by outwitting twenty other equally determined suitors. Rapunzel of the golden hair was pretty tepid stuff in comparison. I often wished that I could believe it, too.

I looked over at Howard. He was sitting there with the glazed look of one who is living elsewhere. There was no doubt in my mind that we had been on the same romantic journey back through the years, but I wasn't sure just where he was pausing to dally. I wanted to know.

"*Two* pennies for your thoughts." I doubled the regular bid on the theory that Howard's thoughts were worth more than the standard price.

"I was thinking," he said thoughtfully, "that, on the whole, summing it up, they've been good years."

This was not a remark with which one could take issue. Yet the statement had a quality of cool appraisal and lacked the warmth with which Howard usually referred to our marriage —or me.

"It's been fun, fascinating in fact; sometimes turbulent but never dull."

I was appeased; these were not bad qualities to have brought to a marriage. And then his further observation jarred me out of my complacency.

". . . and, until recently, successful years—all but one!"

"*I* wasn't aware of it," I said and was surprised that my tone betrayed none of the crashing sounds inside my head.

"Perhaps not. It was some years ago and I didn't like to worry you with it. Besides, by the following year things were better . . . and I guess I thought they'd stay that way."

So this was the way such things happened—calmly, impersonally. It was a nightmare, written by Frederick Lonsdale and set in an English drawing room. Any minute a butler would come in and say "Will Modom have a divorce here or on the terrace . . . and with cream or lemon?"

Occupation: Angel

"But things *haven't* stayed better?" I might as well have it spelled out; get it all at once.

"No. Surely even *you* must be aware of that. The last two years have been, what shall I say, not good . . . not bad? It's logical that they wouldn't be as exciting as the early years, we couldn't expect that. Still, you must admit," he said, and I was shocked at the cheery note in his voice as he summed it up, "that if we should call it quits now only *one* really bad year in twenty-five is a pretty good record."

"Twenty-*seven* years," I said, clinging to reality.

"That's strange," Howard responded, "I'd have sworn it was less. Are you sure it's twenty-seven?"

"Definitely! Maggie's twenty-*five*," I said pointedly.

"There!" A note of triumph was in his voice. "That *proves* I'm right. We started when she was a baby, remember?"

"We started . . . what?" I asked stupidly.

"We started investing in the theater! Haven't you been listening to anything I've said? Really! I always thought Elmer Rice had you in mind when he wrote *Dream Girl*. Maybe I should have asked him how to cope with the problem. I assumed we were talking about the theater." And he pointed an accusing finger at my desk, which certainly corroborated his theory.

"You're absolutely right," I conceded. "And I *had* been thinking about the theater, the end of our twenty-fifth season in it—but then I got to thinking about you, and the day after tomorrow being our twenty-seventh wedding anniversary and how right our marriage has been."

* * *

"How unromantic can you be," he said later, "getting marriage and the theater confused?"

"You've hit upon something there. Somehow, in our case they are entwined; and it's not unromantic, either. They're

Then and Now

both dreams that take an awful lot of time and care and nourishment if they're going to be made to work. And I've never forgotten that in the beginning you considered the theater as a little bit of extra insurance. You didn't really mean what you said about calling it quits on our angeling, did you?"

"I'm not so sure. Maybe we should have sense enough to quit while we're this far ahead." Howard seemed to be thinking aloud rather than arguing. "Things have changed so. Look at your own figures . . . costs up over 600 per cent, returns down; odds on a show succeeding six to one against us and every day, one way or another, more *potential* hits being taken out of the investment market. The risks are too great; the rewards too limited. The whole situation is impossible!"

"That's just it!" Suddenly it all seemed more clear to me. "When things become impossible, they change because they have to. The overgreedy ones eventually will choke themselves. And it's just possible that the 'old days' paid too great a proportionate return to the *backers*. Perhaps we need a new look at the whole thing; a fair re-evaluation. I have a feeling that the theater is on the verge of a complete upheaval, a new look, right now. In fact, it has already started and I should have been aware of it before. Some of the old-time realism is being abandoned because we can't afford it. And perhaps intellectually we never could afford it. Look at *A Man For All Seasons:* one part props to nine parts imagination. One could almost see the words take form. And Dick Rodgers did it with *No Strings*. Do you remember the wonderful walk the lovers took through Paris—on a practically bare stage? I followed every step they took; and they stopped at *my* favorite spots on the streets that *I* knew, so that it was infinitely personal. And the musicians were freed from their pit; mood and music and action were all one.

"I think the theater of ancient Greece must have had some of this essence; and it lasted for hundreds of years. . . . We

Occupation: Angel

may be on the very eve of a new renaissance. We can't stop now."

Howard did not need much persuasion but he pointed out that it was my enthusiasm, not my logic, that he was following.

"If we were to choose just as carefully—I don't necessarily mean the conservative plays, we've never done that," he hastened to reassure me. "And perhaps invest half as much money, in half as many plays . . ."

"Nothing *really* changes very much, does it?" I interrupted. "We actually said all this a long time ago. And in a broader sense, so did Thornton Wilder. Don't you hear the echo: 'Oh, oh, oh! Here it is six o'clock and the master not home yet'?"

APPENDIX

Random Notes to Investors

If a producer seeks investment from more than a handful (exact number unspecified) of people known to him or from investors outside of New York State, this approach comes under the heading of a public solicitation and he then must register under the Securities Act. As a source of protection for the backer, the Securities and Exchange Commission requires that the investor receive a prospectus approved by it, so that from the prospectus the investor will be able to get all the information he needs. However, it is helpful if the backer knows how to interpret and evaluate this data.

1. Almost every play on Broadway these days is produced by a Limited Partnership organized under the laws of New York State. Most producers use the standard form but some-

Occupation: Angel

times certain provisions of the standard form are changed to suit the individual purposes of the producer concerned. Customarily, at the end of the form there are set forth particular facts pertaining to the specific production as well as any variations from the usual provisions. In effect, the Limited Partnership provides that after the repayment of the initial investment and the establishment of a reserve (which is variable), all subsequent profits are divided between the general partners and the limited partners (or backers). Frequently the agreement will provide for what is called an involuntary overcall. That is to say, if the capital is not sufficient to meet production or operating costs, the backers may be called upon to pay a specified additional percentage of their original investment. This overcall must be paid back before there can be any distribution of the profits. These profits usually are divided fifty-fifty between the producer and the backers but sometimes, when production expenses are exceptionally high, producers have been known to split sixty-forty to the investors' benefit; or, if the play is a very attractive investment and the producer feels that there will be a big demand for investment in it, he may give only forty per cent of the profits to his limited partners and retain sixty per cent for himself.

2. In addition to the operating profits from the run of the play, the investors share with the producer in the profits of road companies and British companies produced, co-produced or leased by the American producer. Also, under the customary terms, if the play runs for twenty-one performances in New York the producer becomes entitled to a forty per cent share of what the author receives from the sales of motion picture, television and radio rights; to stock and amateur performances; and sometimes, in the case of musicals, to the sale of show album records. All of such proceeds become part of the income from the production and are divided between the producers and the investors. However, in the case of a musical it should

Appendix

be noted that the term "subsidiary rights" does not include the proceeds derived by the authors of the music and lyrics of the play from the publication, mechanical reproduction, synchronization and small performing rights in the separate musical numbers of the play.

3. The prospective angel should read carefully the provisions at the end of the Limited Partnership Agreement to ascertain if (a) the producer is getting the customary rights from the author, or (b) if the producer is giving the customary distribution to the limited partners (or backers), and (c) whether or not percentages of the gross receipts or of the profits are being paid to stars, directors, producers or other personnel and if so, how large a proportion.

a. The motion picture rights may already have been sold by the author so that the producer either does not share in the motion picture rights at all or must yield a percentage thereof (usually one-third) to the motion picture company.

b. Sometimes the producer limits the extent to which the investors may share in the subsidiary rights. Thus, in a current Limited Partnership Agreement, the producer is not allowing the investors to share in any of the subsidiary rights except the possible motion picture proceeds, thereby considerably curtailing prospective income to the backers.

c. If a star (or stars), producer, director, choreographer, designer or other personnel gets a percentage of the weekly gross or of the profits, there will be that much less to distribute to the investors. Usually the producer is careful to protect the rights of the investors, but in the hands of a stupid, overly-optimistic (or dishonest) producer it would be perfectly possible for so high a percentage of the weekly gross to be pledged to author, producer, director, scenic designer, choreographer and star or stars that at the end of a year's seemingly successful run of the play the backers might sustain an actual loss, or receive a miniscule profit on their highly speculative invest-

Occupation: Angel

ment.

4. If, in the initial solicitation of interest for a proposed play, the producer should state that the capacity gross at the theater under consideration is approximately forty-five thousand dollars and that the play will break-even at twenty-five thousand, the neophyte backer must not assume that the difference between the gross figure and the break-even figure represents profit. Theater rental as well as the author's royalties will go up in proportion to the gross intake; so will the profits of anyone who has a contract based upon a percentage of either gross receipts, profits, or both. So that what might, at a cursory glance, look like a possible profit at capacity business of twenty thousand dollars a week, more realistically could yield from one half to one quarter that much.

The cost of producing a non-musical play on Broadway usually ranges from ninety thousand to one hundred and sixty thousand; musicals are likely to run from a low of three hundred thousand up to six hundred and fifty thousand and more.

An estimated production budget is at best an experienced guess, and actual costs may be less or considerably more. In cases where the expenses have exceeded the capitalization (including a possible overcall), the producer is personally responsible for the additional money needed.

ESTIMATED PRODUCTION BUDGET ON A NON-MUSICAL ONE-SET PLAY
(without overall) . . . $150,000.00

Artistic Production		
Scenery—Design	$2,500	
" Building and Painting	18,000	
Costumes—Design and Purchase	4,000	
Props—Purchase and Rental	6,000	
Electrical and Sound Equipment	6,500	
Director's Fee	5,000	($42,000)
Rehearsal Expenses		
Salaries: Actors and Understudies	$5,000	
Company Crew	4,000	
Stage Hands Take-In and Hang	4,500	
Stage Managers	2,200	
Wardrobe Mistress	300	
Production Assistant	400	
Theater Rent and Expenses	1,800	
Scripts and Miscellaneous	1,800	($20,000)
Advance Advertising		
Newspaper Ads	$7,500	
Photographs and Signs	1,500	
Printing, Promotion and Display	3,000	
Press Agent—Salary and Expenses	2,000	($14,000)
Administration		
Legal and Auditing Fees	$3,000	
Company and General Managers	3,500	
Office Expense	2,500	
Railroad, Transportation, Cartage	3,000	
Payroll Taxes	3,000	
Insurance, Blue Cross, etc.	2,500	
Miscellaneous on Pre-production	1,500	($19,000)
Advances and Bonds		
Advance to Authors	$1,000	
Actors' Equity	18,000	
Theatrical Unions (ATPAM and IATSE)	3,000	
Theater	6,000	($28,000)
Reserve for Tryout Contingencies	$12,000	
Reserve for Sinking Fund, Working Capital	15,000	($27,000)
Total Capitalization		$150,000.00

ESTIMATED PRODUCTION BUDGET ON A MUSICAL
(without overcall) $500,000.00

Artistic Production
    Scenery—Design and Building $80,000.00
    Costumes—Design and Execution   75,000.00
    Furniture and Props   7,000.00
    Draperies   7,500.00
    Electrical and Sound Equipment   8,000.00
    Miscellaneous   1,500.00   ($179,000.00)

Rehearsal Expenses
    Salaries: Actors and Understudies $24,000.00
            Director   5,000.00
            Musical Director   3,500.00
            Choreographer   3,000.00
            Dance Arranger and Piano   2,500.00
            Choral Director and Piano   1,500.00
            General and Company M'gr   6,500.00
            Stage Managers   4,000.00
            Company Crew   7,000.00
            Musicians   7,500.00
            Wardrobe   300.00
    Arrangement and Orchestrations   30,000.00
    Theater Rent and Expenses   2,700.00
    Stage Hands—Set-Up and Hang
      (Out of town)   6,500.00
    Stage Hands—Set-Up and Hang
      (In N.Y.)   6,500.00
    Hauling   5,000.00
    Scripts and Miscellaneous   1,500.00   ($117,000.00)

Advance Advertising		
Newspaper Ads	$16,000.00	
Photographs, Signs, Printing, etc.	8,500.00	
Press Agent and Promotion Expenses	3,500.00	($28,000.00)
Administration		
Legal Fees, Auditing and Expenses	$9,500.00	
Office	2,400.00	
Transportation	2,600.00	
Payroll Taxes, Insurance, etc.	7,500.00	
Auditions and Pre-opening Expenses	5,500.00	
Travel and Expense Allowance (Out of town)	14,500.00	
Auditing, Secretary, etc.	2,500.00	
Miscellaneous on Pre-production	1,500.00	($46,000.00)
Advances and Bonds		
Advance to Authors, Composer, Lyricist	$5,000.00	
Actors' Equity	29,000.00	
Unions (ATPAM, IATSE and 802)	8,000.00	
Theater Guarantees (Out of town)	15,000.00	
Theater Guarantee (N.Y.)	12,000.00	($69,000.00)
Reserve for Tryout Contingencies	$21,000.00	
Reserve for Sinking Fund and Working Capital	40,000.00	
Total Capitalization		$500,000.00

AUDITOR'S WEEKLY STATEMENT ON CURRENT, OUTSTANDING COMEDY HIT

Box Office Receipts	$42,299.79
Less: Theater Share	11,574.95
Company Share	$30,724.84
Expenses	
Salaries	
Cast	$9,129.98
Crew	438.96
Stage Manager	268.75
Company and General Managers	625.00
Press Agent and Promotion	405.00
Wardrobe and Dressers	360.00
Extra Stagehands	130.36
Royalty	
Author	4,412.01
Director	1,323.60
Designer	125.00
Producer's Fee	423.00

Publicity	
Share of Newspapers	708.11
Photos and Signs	271.36
Printing and Promotion	1.95
Press Expense	40.76
Reserve for Future Advertising	300.00
Departmental	
Electrical	11.24
Props	167.20
Costume Replacement	370.80
Carpenter	50.00
Rentals	149.35
Office Expense	350.00
Auditing	75.00
Payroll Taxes	836.86
Insurance	150.00
N.Y. City Excise Tax	128.19
League Dues	50.00
Share of Box Office Staff	294.03
Miscellaneous	157.75
Total Expenses for Week	**$21.754.26**
Running Profit for Week	**$ 8,970.58**

(Share of profit to be divided among investors after initial cost of production has been repaid ......$4,485.00)

AUDITOR'S WEEKLY STATEMENT ON CURRENT, OUTSTANDING MUSICAL HIT

Box Office Receipts	$65,768.04
Less: Theater Share	16,942.01
Company Share	$30,724.84
Expenses	
Salaries	
Cast	$10,917.29
Stage Managers	550.00
Crew	1,975.00
Wardrobe	740.30
Company Manager	400.00
Press Agent	300.00
Orchestra	535.00
Extra Salaries (Company Share)	
Musicians	2,977.23
Treasurers	314.24
Royalties	9,313.10
Publicity	
Newspapers	1,279.33
Photos, Printing, etc.	74.41
Billboards and Signs	1,454.00
Press Agent	109.68

Departmental	
Wardrobe	642.82
Electrical	260.76
Property	40.90
Other	77.48
Rentals	602.63
Other Expenses	
Office Charges	350.00
Insurance	275.00
Auditor	90.00
League Dues	50.00
Miscellaneous	19.11
Payroll Taxes	267.23
NYC Gross Receipts Tax	203.86
Air Conditioning	648.53
Total Expenses for Week	$34,467.90
Running Profit for Week	$14,358.13

(Share of profit to be divided among investors after initial cost of production has been repaid ...... $7,179.06)

RECORD OF BROADWAY PRODUCTIONS

Season	Plays	Musicals	Revivals	Total
1901–1902	49	21	20	90
1906–1907	67	34	28	129
1911–1912	85	39	16	140
1916–1917	85	25	16	126
1921–1922	142	37	15	194
1926–1927	188	49	26	263
1931–1932	146	27	34	207
1936–1937	94	11	13	118
1941–1942	58	16	9	83
1946–1947	48	14	17	79
1951–1952	44	9	19	72
1956–1957	37	10	15	62
1961–1962	32	17	1	50

BREAKDOWN ON TYPE OF PRODUCTIONS
Season of 1961–62

Plays (32)
    New   32
    Revivals   0
(Of this number 9 were adaptations, 8 imports.)

   Musicals (18)
    New   17
    Revivals   1
(Of this number 9 were adaptations, 1 import.)

Total (50)